Interpenetrating
Polymer Networks
and Related Materials

Interpenetrating Polymer Networks and Related Materials

L.H. Sperling

Lehigh University
Bethlehem, Pennsylvania

Plenum Press · New York and London

Library of Congress Cataloging in Publication Data

Sperling, Leslie Howard, 1932-
   Interpenetrating polymer networks and related materials.

   Includes index.
   1. Graft copolymers. I. Title.
QD382.G7S63   547.8'4   80-20344
ISBN 0-306-40539-3

© 1981 Plenum Press, New York
A Division of Plenum Publishing Corporation
227 West 17th Street, New York, N.Y. 10011

Printed in the United States of America

To my younger daughter, *Sheri,* who
composes "Music to Make Polymers By,"
and my older daughter, *Reisa,* who
dances to it.

PREFACE

To the surprise of practically no one, research and engineering on multi-polymer materials has steadily increased through the 1960s and 1970s. More and more people are remarking that we are running out of new monomers to polymerize, and that the improved polymers of the future will depend heavily on synergistic combinations of existing materials.

In the era of the mid-1960s, three distinct multipolymer combinations were recognized: polymer blends, grafts, and blocks. Although inter-penetrating polymer networks, IPNs, were prepared very early in polymer history, and already named by Millar in 1960, they played a relatively low-key role in polymer research developments until the late 1960s and 1970s.

I would prefer to consider the IPNs as a subdivision of the graft copolymers. Yet the unique topology of the IPNs imparts properties not easily obtainable without the presence of crosslinking.

One of the objectives of this book is to point out the wealth of work done on IPNs or closely related materials. Since many papers and patents actually concerned with IPNs are not so designated, this literature is significantly larger than first imagined. It may also be that many authors will meet each other for the first time on these pages and realize that they are working on a common topology.

The number of applications suggested in the patent literature is large—and growing. Included are impact-resistant plastics, ion exchange resins, noise-damping materials, a type of thermoplastic elastomer, and many more.

On the research side, the concept of an IPN titillates the imagination. In what sense do they interpenetrate? In how many significantly different ways can two networks be placed in juxtaposition? The several dozen variations already reported are described and compared here. I hope that this book will serve to encourage the search for yet different modes of combination.

I wish to thank Dr. Kurt C. Frisch, Dr. Harry L. Frisch, Dr. Daniel Klempner, and Dr. Yuri S. Lipatov, among many others, for their help in preparing this book. If not for their researches, the field of IPNs would be much diminished. My collaborators at Lehigh University, Dr. David A. Thomas and Dr. John A. Manson, provided many of the ideas, as well as results, that are incorporated in this book. My students, too numerous to name here, labored long but fruitfully to garner an important part of the data included in this book.

I wish to thank Lehigh's science librarians, Ms. Sharon L. Siegler, Ms. Christine L. Roysdon, and Ms. Linda Khatri, who patiently helped with the various aspects of literature searching, and Miss Andrea Weiss, who did much of the photography.

Lastly, the secretarial work of Ms. Karen Christman, Mrs. Carole Willis, Mrs. Diane DelPriore, and Ms. Jone Svirzofsky must be mentioned, for they alone were able to translate my notes and scribbles into ordinary typewritten pages.

L. H. Sperling

CONTENTS

AN INTRODUCTION TO POLYMER NETWORKS AND IPNs

An interpenetrating polymer network, IPN, can be defined as a combination of two polymers in network form, at least one of which is synthesized and/or crosslinked in the immediate presence of the other. An IPN can be distinguished from simple polymer blends, blocks, and grafts in two ways: (1) An IPN swells, but does not dissolve in solvents, and (2) creep and flow are suppressed.

Because the great wealth of polymer literature fills many volumes each year, an introductory chapter cannot possibly encompass all of the ideas and concepts, let alone the experimental facts. On the other hand, it is obvious that polymer scientists and engineers have barely scratched the surface of what is rapidly becoming one of the major areas of technological learning. The reader is referred to the Suggested Reading at the end of the chapter for a more complete exposition of polymer science. This introductory chapter will be devoted to the development of a few concepts needed to understand the subject of interpenetrating polymer networks, and a brief perusal of the historical development of IPNs and related materials.

1.1. ON THE NATURE OF A CROSSLINK

Polymers may be synthesized as linear, branched, or crosslinked entities, or a mixture containing all three. In many cases, the complete gamut between a strictly linear chain and a densely crosslinked network can be covered systematically. As extreme examples, high-density polyethylene serves as the model for a strictly linear polymer, and the diamond as the most densely crosslinked polymer imaginable.

A chemical crosslink may be defined as a covalent junction with a functionality greater than two, the chain segments of which generally extend to other crosslink sites, thus forming a network. The theory of gelation[1] defines a network as the point where the molecular weight becomes infinite. Figure 1.1. illustrates the nature of a crosslink. Such a polymer is insoluble (but can be highly swellable), and technically consists of one macroscopic

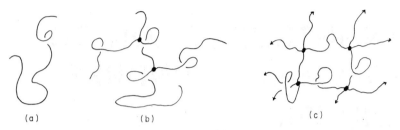

Figure 1.1. The concept of chemical crosslinking. (a) a linear chain, (b) a branched chain, and (c) a crosslinked network. Arrows indicate chains going to further crosslink sites.

molecule.* On the submicroscopic level, a branch point is chemically identical to a crosslink, except that the chains generally terminate after a finite number of further branches. In the simplest case, the number of further branches is zero. Thus a branched polymer may have a high molecular weight and a broad-molecular-weight distribution, but by and large it remains soluble.

A word should be said about physical crosslinks. A physical crosslink is a bond of a physical nature, which joins two or more chains together. Three types of physical crosslinks can be readily identified. They arise from (1) crystalline portions of a semicrystalline polymer, (2) the glassy or crystalline portion of a block copolymer, and (3) the ionic portion of an ionomer. Frequently, the physical crosslink forces can be reduced or overcome by raising the temperature. Thus such materials are thermoplastic at high temperatures, and thermoset at room temperature.

While most of the materials considered herein are crosslinked chemically, several sections are devoted to physically crosslinked systems. This becomes all the more important when one notes that several recent patents and papers utilize the latter to synthesize IPN-like materials having a thermoplastic nature.[2]

Entanglements between chains also serve as a type of crosslink. In a linear or branched polymer, entanglements can slip or move, and so are very impermanent. However, chemical (or physical) crosslinking limits their motion, and increases their effect on bulk properties. At this time, the phantom network theory[3] is calling into question the reality of entanglements. While a monograph such as this cannot of itself resolve the controversy, some of the properties of interpenetrating polymer networks described in later chapters bear on the problem.

---

* Barring minor impurities, an ordinary rubber band consists of only one molecule. This fits both portions of the definition of a molecule. (1) A Maxwell demon can traverse the entire system by stepping on adjacent covalent bonds, and (2) the entire rubber band is the smallest entity having the properties of rubber bands. Breaking the rubber band creates a new material, for then it cannot bind things together in the ordinary manner.

1.2. DEFINITION OF AN INTERPENETRATING POLYMER NETWORK

In its broadest definition, an interpenetrating polymer network, IPN, is any material containing two polymers, each in network form.[4-13] A practical restriction requires that the two polymers have been synthesized and/or crosslinked in the immediate presence of each other. Two types of IPNs are illustrated in Figure 1.2.[14] The sequential IPN begins with the synthesis of a crosslinked polymer I. Monomer II, plus its own crosslinker and initiator, are swollen into polymer I, and polymerized *in situ*. Simultaneous interpenetrating networks, SINs, begin with a mutual solution of both monomers and their respective crosslinkers, which are then polymerized simultaneously by noninterfering modes, such as stepwise and chain polymerizations.

A third mode of IPN synthesis takes two latexes of linear polymers, mixes and coagulates them, and crosslinks both components simultaneously. The product is called an interpenetrating elastomeric network, IEN. There are, in fact, many different ways that an IPN can be prepared; each yields a distinctive topology.

The term "interpenetrating polymer network" was coined before the full consequences of phase separation were realized. Molecular

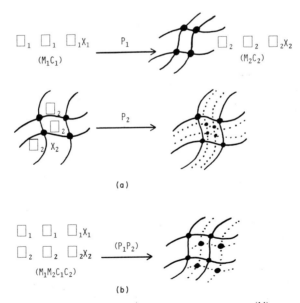

(a)

(b)

Figure 1.2. A schematic comparison of (a) IPN and (b) SIN synthesis.[14] Solid lines represent network I, dotted lines network II. The heavy dots represent crosslink sites.

interpenetration occurs only in the case of total mutual solubility; however, most IPNs phase separate to a greater or lesser extent. Thus molecular interpenetration may be restricted or shared with supermolecular levels of interpenetration. In some cases, true molecular interpenetration is thought to take place only at the phase boundaries.

Given that the synthetic mode yields two networks, the extent of continuity of each network needs to be examined. If both networks are continuous throughout the sample, and the material is phase separated, the phases must interpenetrate in some way. Thus some IPN compositions are thought to contain two continuous phases.[15] The extent of molecular mixing at the phase boundaries has aroused significant interest lately.[16]

Figure 1.3 illustrates some polymer blend and IPN structures.[9] Structure (a), lacking any chemical bonds between the two polymers, is called a polymer blend. The graft copolymer in schematic structure (b) shows a side chain of polymer II emanating from the side of polymer I. The position of the graft linkage is often random, and not restricted to one side chain (see below). In a block copolymer, the chains are joined end-to-end. Structure (c) illustrates a triblock copolymer, the basis for the thermoplastic elastomers. When only one of the polymers is crosslinked, the product is called a semi-IPN structure (d). If the polymerizations are sequential in time, four semi-IPNs may be distinguished. If polymer I is crosslinked and polymer II is linear, the product is called a semi-IPN of the first kind, or semi-I. If polymer I is linear and polymer II is crosslinked, a semi-II results. The remaining two compositions are materialized by inverting the order of polymerization. For

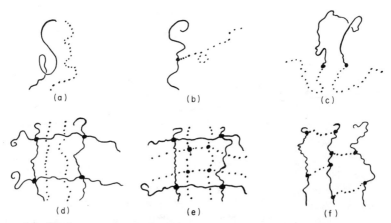

Figure 1.3. Simple two-polymer combinations. (a) illustrates a polymer blend, without covalent linkages. (b) is a graft copolymer, (c) is a block copolymer, (d) illustrates a semi-IPN, (e) a full IPN, and (f) designates an AB-crosslinked polymer. Structure (f) requires two polymers to make one network, while (e) has two independent networks.[9]

simultaneous polymerizations, of course, only two semi-IPNs or semi-SINs may be distinguished.*

Structure (e) of Figure 1.3 illustrates an IPN, with both polymers crosslinked. While grafts between networks I and II may occur to greater or lesser extents, the IPN topology may be said to exist if the deliberately introduced crosslink sites outnumber the accidently introduced graft sites. Structure (f), an AB-crosslinked polymer, requires two polymers to form one network.

It should be pointed out that while many block copolymer systems, structure (c), are substantially 100% actually in block form, many of the most important graft copolymers such as ABS and HiPS are only slightly grafted, being some 80–95% composed of the two homopolymers.

1.3. HISTORICAL DEVELOPMENT OF IPNs

As in many other areas of scientific and engineering endeavor, it is difficult to pinpoint an exact time of origin for the ideas leading to IPNs. Clearly the discovery of a synthetic means of crosslinking polymers belongs in this chain of events. Since no other logical time of origin appears better, Table 1.1 begins with Goodyear's work on vulcanization, or crosslinking, of rubber, and includes the development of polymer blends, grafts, and blocks.

Since the chainlike characteristics of polymeric materials were not understood until H. Staudinger's work beginning in 1920, systematic research on polymer topology could hardly have preceded his efforts. However, within a decade, Ostromislensky's patent clearly shows an understanding of graft copolymer structure. During the period of the 1940s and 1950s, great advances were made with polymer blends, grafts, and block copolymers.

These structures are illustrated in Figure 1.3. Through the leadership of Amos and others, polymer blends and grafts found uses as rubber-toughened plastics, which include high-impact polystyrene (HiPS) and acrylonitrile–butadiene–styrene (ABS) plastics. As further illustrated in Table 1.1, block copolymers containing a water-soluble block and an oil soluble block became important as surfactants through the work of Lunsted, while other block copolymers, composed of elastomer and plastic blocks, were useful as thermoplastic elastomers.

J. J. P. Staudinger, H. Staudinger's son, began his efforts during this time, applying for the first IPN patent in 1941, which was issued ten years

---

* In the series of papers by K. C. Frisch, H. L. Frisch, D. Klempner, and co-workers, the semi-SINs are called pseudo-SINs.

Table 1.1. History of IPNs and Related Materials

Event	First investigators	Year	Ref.^{a}
Vulcanization of rubber	Goodyear	1844	a
IPN type structure	Aylsworth	1914	b
Polymer structure elucidated	Staudinger	1920	c
Graft copolymers	Ostromislensky	1927	d
Interpenetrating polymer networks	Staudinger and Hutchinson	1951	e
Block copolymers	Dunn and Melville	1952	f
HiPS and ABS	Amos, McCurdy, and McIntire	1954	g
Block copolymer surfactants	Lunsted	1954	h
Homo-IPNs	Millar	1960	i
Thermoplastic elastomers	Holden and Milkovich	1966	j
AB crosslinked copolymers	Bamford, Dyson, and Eastmond	1967	k
Sequential IPNs	Sperling and Friedman	1969	l
Latex IENs	Frisch, Klempner, and Frisch	1969	m
Simultaneous interpenetrating networks	Sperling and Arnts	1971	n
IPN nomenclature	Sperling	1974	o
Thermoplastic IPNs	Davison and Gergen	1977	p

^{a} References:
a. C. Goodyear, U.S. Pat. 3,633 (1844).
b. J. W. Aylsworth, U.S. Pat. 1,111,284 (1914).
c. H. Staudinger, *Ber. Dtsch. Chem. Ges.* **53**, 1073 (1920).
d. I. Ostromislensky, U.S. Pat. 1,613,673 (1927).
e. J. J. P. Staudinger and H. M. Hutchinson, U.S. Pat. 2,539,377 (1951).
f. A. S. Dunn and H. W. Melville, *Nature* **169**, 699 (1952).
g. J. L. Amos, J. L. McCurdy, and O. R. McIntire, U.S. Pat. 2,694, 692 (1954).
h. L. G. Lunsted, U.S. Pat. 2,674,619 (1954).
i. J. R. Millar, *J. Chem. Soc.*, 1311 (1960).
j. G. Holden and R. Milkovich, U.S. Pat. 3,265,765 (1966).
k. C. H. Bamford, R. W. Dyson, and G. C. Eastmond, *J. Polym. Sci.*, **16C**, 2425 (1967).
l. L. H. Sperling and D. W. Friedman, *J. Polym. Sci. A-2*, **7**, 425 (1969).
m. H. L. Frisch, D. Klempner, and K. C. Frisch, *Polym. Lett.* **7**, 775 (1969).
n. L. H. Sperling and R. R. Arnts, *J. Appl. Polym. Sci.* **15**, 2317 (1971).
o. L. H. Sperling, in *Recent Advances in Polymer Blends, Grafts, and Blocks*, L. H. Sperling, ed., Plenum, New York (1974).
p. S. Davison and W. P. Gergen, U.S. Pat. 4,041,103 (1977).

later on the manufacture of smooth-surfaced, transparent plastics. The first use of the term "interpenetrating polymer network" was by Millar in 1960, who also made the first serious scientific study of IPNs. Millar used polystyrene/polystyrene IPNs as models for ion-exchange resin matrices.[17,19]

Unlike most other areas of science and technology, the IPN topology apparently was discovered and rediscovered several times. While Millar was aware of the earliest works, both the Sperling and Frisch teams later arrived at the idea independently but at nearly the same time. The first IPN known, however, was invented by Aylsworth[20] in 1914. This was a mixture of natural rubber, sulfur, and partly reacted phenol–formaldehyde resins. On curing, an IPN is formed. Of course, the patent does not use the word "polymer," or any modern polymer concepts.

The Frisch team was originally composed of two brothers, Harry, at the State University of New York at Albany, and Kurt, at the University of Detroit. Daniel Klempner was at that time Harry Frisch's student. Later he became a faculty member at Detroit and a full-fledged team member. H. Frisch had long been interested in catenanes, which consist of interlocking ring structures, physically bound together as illustrated in Figure 1.4.[21,22] The Frisch team conceived of the IPNs as the macromolecular analog of the catenanes. It should be pointed out that K. Frisch is one of the world's top polyurethane scientists. Because of his interest in polyurethanes, one component of the Frisch team IPNs nearly always consists of a polyurethane.

Sperling was originally interested in producing finely divided polyblends without the need for heavy mechanical mixing equipment. The dual-network idea was fancied as a means of suppressing gross phase separation. Shortly after Sperling began his IPN studies, he was joined by David Thomas, a metallurgist. Thomas brought important electron microscopic and mechanical behavior concepts into the program, while Sperling's contribution was a knowledge of viscoelasticity and the vagaries of network topology and phase structures.

While all of the original thought patterns cannot be described here, two more people must be mentioned. Yury Lipatov of the Academy of Sciences of the Ukranian SSR, Kiev, USSR became interested in the interphase regions of IPNs caused by their thermodynamic incompatibility.[5,16] Dr. Lipatov was the first to consider the IPNs as polymer/polymer composites, where the second network constitutes a filler to the first formed network.

Guy Meyer, of Louis Pasteur University, Strasbourg, France, came to IPNs through a consideration of semi-IPNs as adhesives.[23] In addition to many scientific papers on IPNs and related materials contributed by numerous authors, the burgeoning patent literature describes quite different approaches to IPN chemistry and technology.[24]

John Millar was kind enough to amplify the early history of IPNs.[25] He wrote that the original term for IPNs was "reswollen polymer networks,"

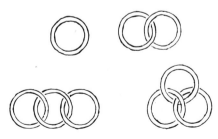

Figure 1.4. Small molecule catenanes. After Millar, unpublished.

but that "interpenetrating polymer networks" was preferred for publication because it was descriptive of the product, not the process. George Solt was a colleague of Millar in Britain and came up with the idea of reswelling a suspension polymer particle to increase bead size. Eventually, this led to anionic/cationic IPNs for ion exchange resins.[18]

A further discussion of Solt's work on anionic/cationic IPNs will be found in Section 7.5. The historical development of phenol–formaldehyde and alkyd-based IPNs and related materials will be found in Section 8.9.

1.4. A RESEARCHER'S APPROACH TO IPNs

According to Cassidy,[26] an interesting research problem must have one of three elements: a new theory, a new instrument, or a new material. This monograph describes the development of a new class of materials, the IPNs. Only the basic elements of the field have yet been explored. For example, the number of papers on IPNs in the period 1977–1979 roughly equals that for the entire preceding period. However, setting down the work already done may inspire yet more research.

The field of IPNs has a special problem relating to nomenclature. Since no standard nomenclature yet exists, and many works on IPNs do not contain the phrase "interpenetrating polymer networks," it is difficult to identify quickly many IPN-related papers and patents. While this work is also incomplete in this respect, many new workers in the field are included.

In brief, the monograph will proceed from the basic to the applied. Since the IPNs are members of the larger "polyblend" field[9,13] elements of polymer blends will be introduced next, followed by a proposed nomenclature. Then the subfield of homo-IPNs will be taken up, as model materials. The main theme of this work will be found in Chapters 5–8.

SUGGESTED READING. BIBLIOGRAPHY OF SELECTED POLYMER BOOKS

F. W. Billmeyer, Jr., *Textbook of Polymer Science*, 2nd ed., Wiley-Interscience, New York (1971).

F. A. Bovey and F. H. Winslow, eds., *Macromolecules: An Introduction to Polymer Science*, Academic, New York (1979).

J. Brandrup and E. H. Immergut, eds., with W. McDowell, *Polymer Handbook*, 2nd ed., Wiley-Interscience, New York (1975).

A. J. Chompff and S. Newman, *Polymer Networks—Structure and Mechanical Properties*, Plenum, New York (1971).

J. K. Craver and R. W. Tess, eds., *Applied Polymer Science*, Organic and Coating Plastics Chemistry, American Chemical Society, Washington, D.C. (1975).

H. G. Elias, *New Commercial Polymers, 1965–1975*, Gordon and Breach, New York (1977).

J. D. Ferry, *Viscoelastic Properties of Polymers*, 2nd ed., John Wiley & Sons, New York (1970).

P. J. Flory, *Principles of Polymer Chemistry*, Cornell University, Ithaca, New York (1953).

P. H. Geil, E. Baer, and Y. Wada, eds., *The Solid State of Polymers*, Report of the U.S.–Japan Joint Seminar, Marcel Dekker, New York (1974).

R. Houwink and H. K. DeDecker, eds., *Elasticity, Plasticity and Structure of Matter*, 3rd ed., Cambridge University (1971).

C. A. Hoyser, *Handbook of Plastics and Elastomers*, McGraw-Hill, New York (1975).

R. W. Lenz, *Organic Chemistry of Synthetic High Polymers*, Interscience, New York (1967).

H. Leverne, *Polymer Engineering*, Chemical Engineering Monographs 1, Elsevier, Amsterdam (1975).

P. Meares, *Polymers: Structure and Bulk Properties*, Van Nostrand, New York (1967).

M. L. Miller, *The Structure of Polymers*, Polymer Science Engineering Series, Reinhold, New York (1966).

L. E. Nielsen, *Mechanical Properties of Polymers*, Reinhold, New York (1962).

G. Odian, *Principles of Polymerization*, McGraw-Hill, New York (1970).

E. M. Pearce and J. R. Schaefgen, eds., *Contemporary Topics in Polymer Science*, Plenum, New York (1977).

A. Ravve, *Organic Chemistry of Macromolecules*, Marcel Dekker, New York (1967).

F. Rodriguez, *Principles of Polymer Systems*, McGraw-Hill, New York (1970).

J. Schultz, *Polymer Materials Science*, Prentice-Hall, Englewood Cliffs, New Jersey (1974).

R. B. Seymour, *Introduction to Polymer Chemistry*, McGraw-Hill, New York (1971).

A. V. Tobolsky, *Properties and Structure of Polymers*, John Wiley & Sons, New York (1960).

B. Vollmert, *Polymer Chemistry*, Springer, New York (1973).

D. J. Williams, *Polymer Science and Engineering*, Prentice-Hall, Englewood Cliffs, New Jersey (1971).

REFERENCES

1. P. J. Flory, *Principles of Polymer Chemistry*, Cornell University, Ithaca, New York (1953), Chap. 9.
2. E. N. Kresge, in *Polymer Blends*, Vol. 2, D. R. Paul and S. Newman, eds., Academic, New York, (1978), Chap. 20.
3. P. J. Flory, *Proc. R. Soc. Lond. Ser. A* **351**, 351 (1976).
4. K. Shibayama and Y. Suzuki, *Rubber Chem. Tech.* **40**, 476 (1967).
5. Yu. S. Lipatov and L. M. Sergeeva, *Russ. Chem. Rev.* **45**(1), 63 (1967).
6. D. Klempner, *Angew. Chem.* **90**, 104 (1978).
7. L. H. Sperling, *Encycl. Polym. Sci. Technol. Suppl.* **1**, 288 (1976).
8. L. H. Sperling, *J. Polym. Sci. Macromol. Rev.* **12**, 141 (1977).
9. D. A. Thomas and L. H. Sperling, in *Polymer Blends*, Vol. 2, D. R. Paul and S. Newman, eds., Academic, New York (1978).
10. H. A. J. Battaerd, *J. Polym. Sci.* **49C**, 149 (1975).
11. D. S. Kaplan, *J. Appl. Polym. Sci.* **20**, 2615 (1976).
12. H. L. Frisch, K. C. Frisch, and D. Klempner, *Mod. Plast.* **54**, 76, 84 (1977).
13. J. A. Manson and L. H. Sperling, *Polymer Blends and Composites*, Plenum, New York (1976), Chap. 8.

14. R. E. Touhsaent, D. A. Thomas, and L. H. Sperling, *J. Polym. Sci.* **46**, 175 (1974).
15. L. H. Sperling, *Polym. Eng. Sci.* **16**, 87 (1976).
16. Yu. S. Lipatov, L. M. Sergeeva, L. V. Karabanova, A. Ye. Nesterov, and T. D. Ignatova, *Vysokomol. Soyed.* **A18**(5), 1025 (1976).
17. J. R. Millar, *J. Chem. Soc.*, 1311 (1960).
18. G. S. Solt, Br. Pat. 728, 508 (1955).
19. J. R. Millar, D. G. Smith, and W. E. Marr, *J. Chem. Soc.*, 1789 (1962).
20. J. W. Aylsworth, U.S. Pat. 1, 111, 284 (1914).
21. H. L. Frisch and E. Wasserman, *J. Am. Chem. Soc.* **83**, 3789 (1961).
22. H. L. Frisch and D. Klempner, *Adv. Macromol. Chem.* **2**, 149 (1970).
23. G. C. Meyer and P. Y. Mehrenberger, *Eur. Polym. J.* **13**, 383 (1977).
24. L. H. Sperling, K. B. Ferguson, J. A. Manson, E. M. Corwin, and D. L. Siegfried, *Macromolecules* **9**, 743 (1976).
25. J. R. Millar, personal communications, December 5, 1978 and January 11, 1979.
26. H. G. Cassidy, *Am. Sci.* **54**, 184 (1966).

2

PHASE SEPARATION AND MECHANICAL BEHAVIOR OF MULTICOMPONENT POLYMER SYSTEMS

2.1. INTRODUCTION

The outstanding behavior of multipolymer combinations usually derives from the phase-separated nature of these materials. In fact, polymer blends, blocks, grafts, and IPNs are interesting because of their complex two-phased nature, certainly not in spite of it. Aspects of phase continuity, size of the domains, and molecular mixing at the phase boundaries as well as within the phase structures all contribute to the mechanical behavior patterns of these multicomponent polymer materials.

This chapter will review the relationships among synthetic detail, morphology, and resulting mechanical behavior. While effects on the glass–rubber transition and modulus will be emphasized, aspects of toughness and impact resistance will be touched upon and applications discussed. In order to generalize this critique, polymer I is defined as the first synthesized polymer, and polymer II as the second synthesized polymer. Even when the order of synthesis is immaterial, as in mechanical blends, this notation will prove useful. Since the basis of this chapter lies in the two-phased nature of these materials, it is appropriate to examine first the fundamental reasons underlying phase separation.

A brief bibliography of multicomponent polymer works is given in the Suggested Reading at the end of this chapter.

2.2. POLYMER I/POLYMER II INCOMPATIBILITY

As stated above, nearly all two-polymer combinations form two immiscible phases. While simple mechanical blends are usually incompatible, it is especially interesting to note that phase separation commonly

11

ensues even when the two polymer chains are joined together to form a single molecule, as in graft or block copolymers. Thus, one end of a molecule is insoluble in the other end, and phases smaller than the macromolecular dimensions are common.

Polymer I/polymer II incompatibility arises from the unusually low entropy of mixing obtained on blending. The Gibbs free energy of mixing, ΔG_M, is given by

$$\Delta G_M = \Delta H_M - T\Delta S_M \tag{2.1}$$

where ΔH_M and ΔS_M represent the enthalpy and entropy of mixing, respectively. Basic thermodynamic treatments have been given by Tompa,[1] Scott,[2] Krause,[3] Inoue et al.[4] Meier,[5] and Donatelli et al.[6]

Equation (2.1) can be written more explicitly in terms of molecular chain structure and interactions:[7-9]

$$\Delta G_M = V(\delta_1 - \delta_2)^2 v_1 v_2 + RT(n_1 \ln v_1 + n_2 \ln v_2) \tag{2.2}$$

where V is the molar volume of the mixture, δ_1 and δ_2 represent the solubility parameters of the two components, n_1 and n_2 represent the number of moles of the two polymers, respectively, and v_1 and v_2 represent their volume fractions. Equation (2.2) is derived in Section 6.5.2. The first term on the right indicates the heat of mixing, and the second term indicates the entropy of mixing.

Demixing occurs because the Gibbs free energy of mixing changes sign, from negative to positive,* as molecular weight increases. Qualitatively, the entropic aspects can be visualized by examining the quasilattice models in Figure 2.1. When the molecules are small and occupy only a single lattice site, Figure 2.1a, all molecules are interchangeable at random. In particular, an ×-type molecule can exchange positions with all other ×-type molecules, and also with all ○-type molecules. Figure 2.1b illustrates the case for two polymer molecules. Three important restrictions on randomness may be noted.

1. All the ×-type mers and all the ○-type mers are now bound with covalent bonds, and the mers within each polymer molecule are no longer interchangeable. For example, two mers of the × type cannot be interchanged without breaking primary covalent bonds. For the same reason, the ○-type mers cannot be interchanged with the ×-type mers any longer.

---

* This assumes that the change in sign for ΔG_M represents the lowest possible free energy states. For example, both the miscible and immiscible states can have ΔG_M negative; of course, that state with the lower value represents the equilibrium state.

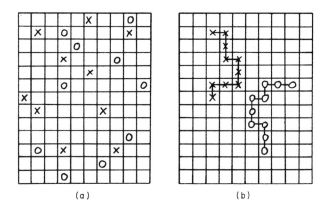

Figure 2.1. Schematic diagram of two quasilattice structures: (a) contains little molecules and (b) contains polymer molecules. The entropy of mixing crosses and circles in (a) is higher than in (b), because more rearrangements are possible.[35]

2. Also because of primary covalent bonding, the $\times$-type and $\bigcirc$-type mers can only move in coordination with their neighbors. Thus an $\times$-type mer can only be one lattice site distance from the adjoining $\times$ mers.

3. Because of the need to maintain chain continuity, one chain cannot physically pass through another, a restatement of the volume exclusion effect[7] as applicable to this problem. In the two-dimensional lattice illustrated in Figure 2.1b, the chains cannot cross. In the real three-dimensional case, positive restrictions also exist regarding placing two chains in juxtaposition.

The ΔH_M term in equation (2.1) is usually positive for nonpolar molecular species, opposing mixing. This is true for big and little molecules alike. While the ΔH_M values for mixing monomer and polymer species obviously differ, the changes after polymerization tend to be modest (and in either direction) in comparison to the entropic changes which decrease dramatically with increasing molecular weight.

Broadly speaking, three cases for mixing thermodynamics can be distinguished:

1. When two species of little molecules are mixed, for example, benzene and carbon tetrachloride, the entropy gained on mixing (times the absolute temperature) is usually larger than the positive heat of mixing, hence ΔG_M is negative, and one phase is formed.

2. For two macromolecular species, $T\Delta S_M$ tends to be smaller than ΔH_M, and two phases result.

3. For the case of one big molecule and one little molecule (a simple polymer solution, for example), the two factors in equation (2.1) are about

the same magnitude.* Thus, total miscibility can be obtained if the solubility parameters equation (2.2) are nearly alike, which results in small ΔH_M terms.

More recent work by Helfand[10-12] and Meier[13] has centered on an understanding of the interfacial region (interphase), giving recognition to the nonclassical concept that the phase boundary is of finite width, sometimes amounting to several tens of angstroms. Even with little molecules, the phase boundary does not consist of a vertical composition cliff, with diffusion and entropic forces causing at least a few angstoms of interphase width. Polymer molecules have contour dimensions of several thousands of angstroms. Since the actual phase dimensions are small the interphase volume can be appreciable, with important consequences for the mechanical behavior of the material.

2.3. POLYMER/POLYMER PHASE DIAGRAMS

While most of the work described in this monograph emphasizes the two-phased nature of IPNs and related materials, it is interesting to explore more deeply the characteristics of phase separation in polymer/polymer systems. Of key importance, McMaster[14,15] showed that most polymer/polymer phase diagrams are expected to exhibit a lower critical solution temperature (LCST). This means that as the temperature is raised the polymer pair becomes less mutually soluble, and phase separates. This effect is not immediately predicted by equations (2.1) and (2.2), which suggest the usual phenomenon of an upper critical solution temperature (UCST).

McMaster[14] demonstrated that LCST behavior arises from free volume effects. In particular, if the pure component thermal expansion coefficients are sufficiently different, LCST behavior becomes more likely. It should be noted again that the normal entropic effects are very small. By using Flory's equation-of-state thermodynamics,[16,17] it can be demonstrated that LCST behavior should generally be expected for high-molecular-weight mixtures. Although the detailed thermodynamic arguments are beyond the scope of this chapter, the theory has been reviewed recently.[18-20]

Several polymer/polymer phase diagrams have now been worked out, and all yielded LCST behavior.[15,20-23] Usually the phase separation curve has been plotted through observation of the cloud point, see Figure 2.2. In

---

* The decrease in the entropic contribution on polymerization of monomer II may be seen in the decrease of n_2 in equation (2.2), for the same total volume.

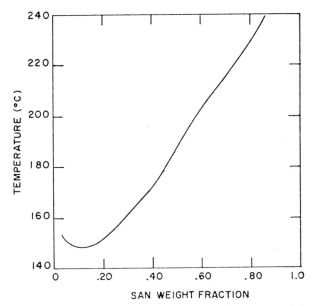

Figure 2.2. Experimental cloud point curve (SAN/PMMA). The region below the line is homogeneous. Phase separation takes place above the line.[15]

order to record the effects in a reasonable time, of course, the system must be above its glass transition temperature.

2.4. MORPHOLOGY OF TWO-PHASED SYSTEMS

Section 2.2 described the thermodynamic reasons causing combinations of two structurally different polymers to phase separate. The actual extent of phase separation (and inversely, the extent of actual intermolecular solution), phase continuity, phase size, and shape all contribute to the mechanical behavior patterns actually observed.

Figure 2.3 shows six graft copolymer and IPN morphologies.[24] In each case, polybutadiene (or an SBR elastomer) served as polymer I, and polystyrene as polymer II. As described in Section 6.3, the unsaturated diene portion was stained with osmium tetroxide, in order to obtain the necessary contrast for these electron micrographs.

The high-impact polystyrene shown in the upper left has polystyrene (a hard, glassy plastic) as the continuous phase, and so would be expected to be hard. The rubber domains have an occluded polystyrene cellular structure, which is really a phase-within-a-phase-within-a-phase. This complexity makes for excellent toughening and impact resistance. By contrast, the

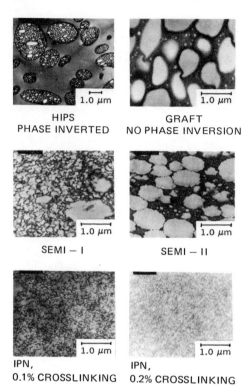

HIPS
PHASE INVERTED

GRAFT
NO PHASE INVERSION

SEMI — I

SEMI — II

IPN,
0.1% CROSSLINKING

IPN,
0.2% CROSSLINKING

Figure 2.3. Six morphologies of graft copolymers, IPNs, and semi-IPNs of SBR/polystyrene. The SBR component is stained dark by osmium tetroxide for transmission electron microscopy. The type of morphology evolved depends on the synthetic detail.[24]

morphology shown in the upper right, prepared without stirring (and hence not phase inverted) but nearly identical chemically to the upper left, has the SBR elastomer as the continuous phase. As a result, the upper right material is much softer, and has poor mechanical properties. Both of these products were prepared from linear polymers, and are best described as graft copolymers. The composition on the upper right, interestingly enough, is very similar to the material prepared by Ostromislensky in 1927, see Table 1.1.

Crosslinking can be introduced in several ways.[24] The middle two morphologies in Figure 2.3 are semi-I and semi-II compositions, left and right, respectively. Crosslinking both polymers results in an IPN, of course. The lower right morphology has smaller domains than the lower left, because the SBR was more densely crosslinked.

Some recent research in the area of IPN-related materials must be noted before proceeding. Eastmond and Smith[25] studied the morphology of AB crosslinked copolymers (ABCPs). Note structure (f) in Figure 1.3. They found that the domain diameters appeared to be proportional to the square root of the degree of polymerization of the domain-forming chains.

2.5. MORE COMPLEX MATERIALS

Since chemists like to idealize chemical compositions, they are often predisposed to assume a regular, simple structure for a material. This is not so for many of the more interesting polymer blend materials. Within the context of the present study, however, only a few examples can be given.

Above, the ABCPs were assumed to follow the idealized structure (f) in Figure 1.3. At crosslinking levels below the gel point, however, the multi-component species present in ABCPs may be considered as linear and nonlinear multiblock copolymers with A branches emanating from junction points such that there are always two A chains linked to one B chain at each junction point. The simplest such structure is the graft copolymer, structure (b) in Figure 1.3. If the B chain is conterminously bound, a triblock copolymer with two A side chains emerges, according to the reasoning of Eastmond and co-workers:[25,26]

$$(2.3)$$

When such a soluble ABCP is mechanically blended with one of its components, unusual morphologies may arise. Eastmond and Phillips[26] observed large spherical domains with an onion-ring-type structure in blends of a soluble polycarbonate/polychloroprene ABCP with poly-carbonate.

Super-high-impact polystyrene can be made from blends of polystyrene with butadiene–styrene block copolymers.[27] As illustrated in Figure 2.4,[28] the morphology of rubber-containing domains is finer than the high-impact polystyrene structure shown in the upper left of Figure 2.3. The subject has been recently reviewed by Aggarwal[28] and Marti and Riess.[29]

2.6. ON DETERMINING PHASE CONTINUITY

As reviewed elsewhere,[28,30–37] block copolymers may have spherical, cylindrical, or alternating lamellar-type morphologies, with either phase continuous depending on the relative proportions of the two blocks. When the lamellar type of morphology is present, both blocks display a degree of phase continuity.

While it is still not possible to predict exactly all possible morphologies of polymer blends, blocks, and grafts,* the following guidelines may be helpful in relating the synthetic detail to phase continuity.[38]

---

* Restricted to the lightly grafted materials, such as ABS, HiPS, and the like. See Section 1.2.

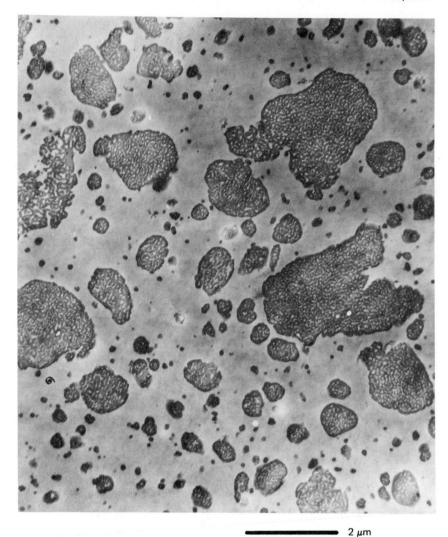

2 µm

Figure 2.4. Morphology of the dispersed phase in the polyblend of 60/40 butadiene/styrene triblock polymer and polystyrene.[28]

1. For simple melt blends, the polymer with the higher concentration or the lower viscosity tends to form the continuous phase. This is illustrated in Figure 2.5.

2. For bulk or solution graft copolymerizations, the polymer first synthesized forms the more continuous phase. Polymer II usually forms cellular domains within polymer I.

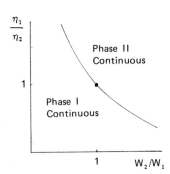

Figure 2.5. Generalized phase continuity diagram, of polymer blends, as controlled by viscosity, η, and weight fraction, w, for polymers 1 and 2. At equal viscosities and weight fractions (marked point), or along idealized curve, some aspects of dual phase continuity may be initiated.

3. Stirring of bulk, suspension, or solution-type graft copolymerizations, especially during the early portion of polymerization II, may cause phase inversion, especially if I is the minority component.

4. For emulsion polymerizations, polymer II, tends to form the continuous phase, after subsequent molding or film formation. In general, the molding of shell–core particulates into macroscopic structures leads to greater continuity of the shell component.

5. For diblock polymers, the relative proportions of the two polymers determine the phase structure and continuity. Midrange compositions have two continuous phases.

6. Crosslinking of either polymer I or polymer II tends to promote phase continuity of the crosslinked component. Materials with both polymers crosslinked (IPNs) tend to develop two continuous phases.

Of course, a significant part of this monograph will be devoted to the morphology and phase continuity of IPNs and related materials.

2.7. MECHANICAL BEHAVIOR OF TWO-PHASED SYSTEMS

One of the great truisms of polymer science says that all amorphous polymers have a glass–rubber transition. The temperature at which this occurs depends primarily on the chemical structure of the polymer. When two polymers form a phase-separated mixture, each retains its glass transition, see Figure 2.6. In general, the transitions may be broadened or shifted by greater or lesser mixing, and of course in the limit of compatibility (mutual solubility), only one transition will be observed. The storage modulus E' is a measure of the stiffness of the material, while the loss modulus E'' is a measure of the amount of energy lost as heat during dynamic testing. The relative value of the two parameters measures the fraction of energy recovered (stored) during a deformation to that lost as heat.

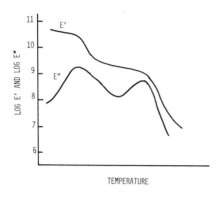

Figure 2.6. Idealized glass–rubber transition behavior of incompatible two-polymer combinations. In general, both glass–rubber transitions will be observed.[35]

Figure 2.7, in a gross oversimplification, illustrates the phenomenon of a rubber ball being bounced on a perfectly elastic floor.† The storage modulus is indicated by the height the ball bounces back, and the loss modulus is the remainder, the energy lost during the collision of the ball with the floor. The formal mathematical relationship is given by

$$E^* = E' + iE'' \tag{2.4}$$

where E^* represents the complex modulus, and i is the square root of minus one. The quantity E'' has an analog in spectroscopy using electromagnetic radiation. For example, using infrared light an absorption peak is noted where the frequency of the light matches the energy required for a chemical group to go to a higher energy state. Likewise, at the correct frequency and temperature, the mechanical waves of energy will be absorbed by the polymer as increased molecular motion, giving rise to a maximum in E''. Hence, experiments to measure E' and E'' are sometimes called dynamic mechanical spectroscopy.

† Even in this world of serious affairs, the reader should take a moment to ponder the initials of the experimenter in Figure 2.7.

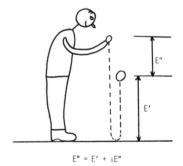

Figure 2.7. Isaac P. Newton makes polymer physics easy. The rubber ball loses some of its energy during a collision with an ideally elastic floor, and its rebound is incomplete.[35]

The loss and storage moduli for mechanial blends of poly(vinyl chloride) with poly(butadiene-*co*-acrylonitrile) are shown[39] as a function of temperature and composition in Figures 2.8–2.10. With increased acrylonitirile content in the elastomer portion, the two polymers become more compatible because of increased polar attractive forces. Figure 2.8 illustrates the glass transition behavior of an incompatible polymer pair, while the modulus changes with temperature in Figures 2.9 and 2.10 indicate greater molecular mixing. In fact, Figure 2.10, with one sharp composition-dependent glass transition, behaves in a manner similar to its isomeric, compositionally equivalent random copolymer.

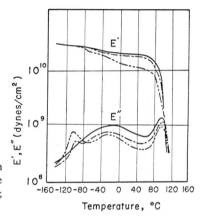

Figure 2.8. Loss and storage modulus of an incompatible PVC/PBD blend. Solid curves are homopolymer PVC; — · —, 100/5 PVC/PBD; — · · —, 100/15 PVC/PBD.

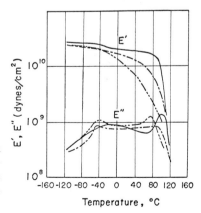

Figure 2.9. Loss and storage modulus for a PVC/P(B-*co*-AN) blend. Addition of 20% acrylonitrile to the elastomer improves compatibility; compare with Figure 2.8. Solid line, 100/0; — · —, 100/15; — · · —, 100/25.[39]

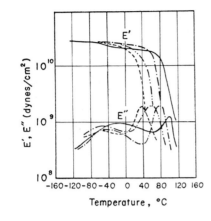

Figure 2.10. Loss and storage modulus for a PVC/P (B-*co*-AN) blend. At the level of 40% acrylonitrile in the elastomer portion, the blend behaves mechanically as if it were one phase. Solid line 100/0; — · —, 100/10; — · · —, 100/25; - - -, 100/50.

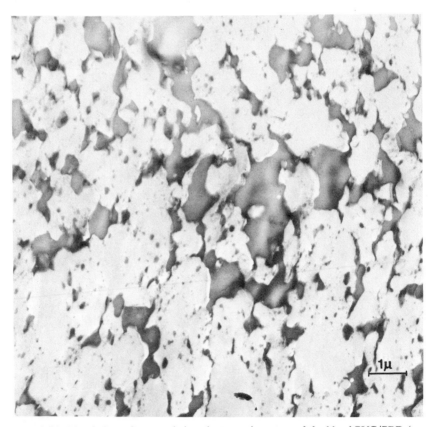

Figure 2.11. Morphology via transmission electron microscopy of the blend PVC/PBD (see Figure 2.8). Sharply defined phase boundaries indicate gross incompatibility.[40]

Electron microscopy on these same materials,[40] Figures 2.11–2.13, show explicitly the phase domain sizes and shapes. With increasing acrylonitrile content, the molecular mixing increases and the phases become smaller and less distinctly demarcated. In Figure 2.13, in fact, the phases are smaller than the molecules! (Phases of less than 100 Å are shown in Figure 2.13, molecular end-to-end distances of 500–800 Å are expected based on dilute solution measurements.)

The two techniques, dynamic mechanical spectroscopy and electron microscopy, yield complementary data. Dynamic mechanical spectroscopy shows more clearly the extent of molecular mixing, while electron microscopy shows the phase domain sizes and shapes. Both techniques contribute to an understanding of phase continuity.

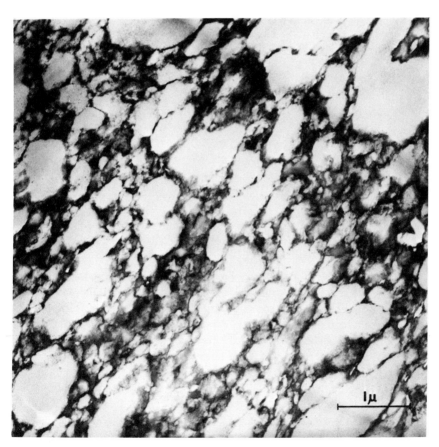

Figure 2.12. Morphology of a PVC/P(B-*co*-AN) blend, containing 20% AN. This semicompatible morphology corresponds to the data in Figure 2.9.[40]

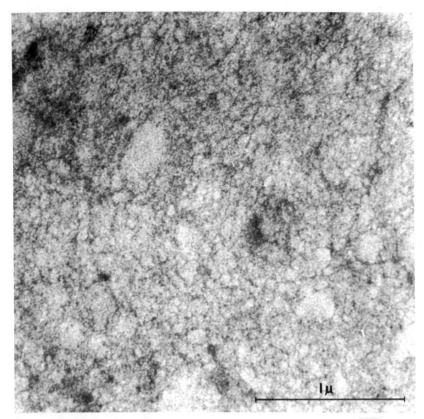

Figure 2.13. Morphology of a PVC/P(B-*co*-AN) blend, containing 40% AN. In contrast to Figure 2.10, electron microscopy reveals a slight extent of phase separation.[40]

Of course, the morphology of the materials is subject to change, by annealing, solvent, or mechanical strain. For example, triblock copolymer thermoplastic elastomers behave quite differently in stress–strain behavior on repeated stress–strain cycling[41] between the first and second cycle, (Figure 2.14). The very significant softening between the first and second cycle is due probably to the destruction of a fibrous glassy polystyrene network phase, which was initially cocontinuous with the elastomer phase.

The effect of block copolymer molecular weight on compatibility and hence mechanical behavior was recently studied by Kraus and Rollman.[42] As the block lengths are decreased, the two polymer transition temperatures become closer, suggesting greater molecular mixing at the interphase, see Figure 2.15. Low-molecular-weight block polymers do not contain compositionally pure phases, but are visualized as retaining a residual

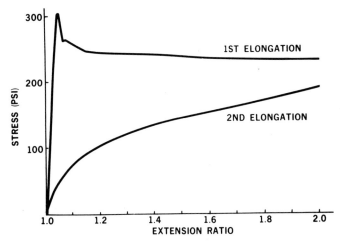

Figure 2.14. First and second stress–strain curves for a block copolymer thermoplastic elastomer. When the polystyrene phase continuity is destroyed during the first elongation, a soft elastomer results.[41]

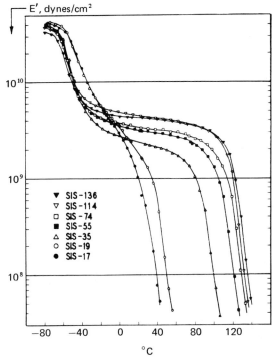

Figure 2.15. Dynamic storage moduli of styrene–isoprene–styrene block copolymers at 35 Hz. Numbers indicate M_ω in thousands. All compositions have about 50% polystyrene.[42]

domain structure in which composition fluctuates between ever-narrowing limits as blocks become shorter. In the limit of very low molecular weight, the mechanical behavior of random copolymers is approached as the system becomes homogeneous.

2.8. APPLICATIONS OF POLYMER BLENDS, GRAFTS, AND BLOCKS

Many properties are involved in determining a material's ultimate usefulness. For homopolymers, the temperature of the glass–rubber transition, presence or absence of crystallinity, crosslinking, molecular weight characteristics, permeability, and toughness all make a contribution. When two different polymers are mixed or joined in some fashion the mode of mixing becomes important as well. Tables 2.1–2.3 describe some of the many actual or proposed applications for two-polymer combinations. In many cases, the patent literature has been referred to, as this is an excellent source of such information. With polymer blends, blocks, and grafts, many new properties and hence uses appeared that were previously unavailable.

Of course, in many of these applications, the mechanical behavior of the materials is paramount. In fact, many such applications, such as impact resistance, arise because of the two-phased nature of the products. In this case, the size of the rubber phase domains and their glass temperatures are

Table 2.1. Applications of Polymer Blends

Ref.[a]	Mode of Combination	Application
a	Rubber-toughened plastics	Impact resistance
b	Saturated rubber/diene rubber	Ozone-crack-growth resistance in tires
c	Bicomponent fibers	Resilient rugs
d	Biconstituent fibers	Reduction of tire flat spotting
e	Multilayer films	Gas transport control
f	Synthetic paper	Strength, water resistance
g	Low mol wt/high mol wt	Processing aid to lower viscosity
h	Wood (Natures' Blend)	Toughness in structural applications

[a] References
a. M. Matsuo, *Jpn. Plast.* **2** (July), 6 (1968).
b. J. F. O'Mahoney, *Rubber Age*, **102** (March), 47 (1970).
c. P. V. Papero, E. Kuba, and L. Roldan, *Text. Res. J.* **37**, 823 (1967).
d. L. Cresentini, U.S. Pat. 3,595,935 (1971).
e. L. M. Thomka and W. J. Schrenk, *Mod. Plast.* **49**(4), 62 (1972).
f. P. W. Morgan, U.S. Pat. 2,999,788 (1961).
g. R. D. Deanin, in *Recent Advances in Polymer Blends, Grafts, and Blocks*, Plenum, New York (1974).
h. R. F. S. Hearmon, in *Mechanical Properties of Wood and Paper*, R. Meredith, ed., Interscience, New York (1953).

critical in determining impact resistance.[1,8] Other applications are suggested because of permeability characteristics, adhesive properties, etc.

In Table 2.1 the bicomponent fibers (which often have half the cylinder as polymer I and the other half as polymer II) are of interest because of the differences in thermal coefficients of expansion. On cooling, the differential rates of contraction cause a certain coiling or curlicue formation in the fibers.

In Table 2.2, the polymer-impregnated wood has reduced porosity, and hence is more water resistant. It is also stronger. The ionomers are included in Table 2.3, because they phase-separate like the block copolymers, and represent the limiting case of polymer I/polymer II having opposite chemical properties. One may consider these products block copolymers where the block consists of a single mer. The thermoplastic elastomers utilize the

Table 2.2. Applications of Graft Copolymers

Ref.[a]	Mode of combination	Application
a	Rubber-toughened plastics	Impact resistance
b	Desalination membranes	Creep reduction
c	Wash and wear finishes	Wrinkle control
d	Immobilized enzymes	Long-lived catalytic surfaces
e	Coatings and adhesives	Toughness
f	Polymer-impregnated wood	Durable knife handles

[a] References:
  a. J. L. Amos, *Polym. Eng. Sci.* **14**, 1 (1974).
  b. H. B. Hopfenberg, F. Kimura, P. T. Rigney, and V. Stannett, *J. Polym. Sci.* **28C**, 243 (1969).
  c F. Rodriguez, *Principles of Polymer Systems*, McGraw-Hill, New York (1970).
  d. M. Charles, R. W. Coughlin, E. K. Paruchuri, B. R. Allen, and F. X. Hasselberger, *Biotech. Bioeng.* **17**, 203 (1975).
  e. C. R. Martens, ed., *Technology of Paints, Varnishes, and Lacquers*, Reinhold, New York (1968).
  f. J. A. Duran and J. A. Meyer, *Wood Sci. Technol.* **6**, 59 (1972).

Table 2.3. Applications of Block Copolymers

Ref.[a]	Mode of combination	Application
a	Plastic/rubber/plastic	Thermoplastic elastomers (no curing required)
b	Water soluble/water insoluble	Surface-active agents
c	Crystalline/rubbery multiblock	Elastomeric fibers (Spandex® type)
d	Crystalline (log T_g)/crystalline	Tough plastics (Polyallomer® type)
e	Ionomers	Tough plastics (Surlyn®)

[a] References:
  a. J. F. Beecher, L. Marker, R. D. Bradford, and S. L. Aggarwal, *J. Polym. Sci.* **26C**, 117 (1969).
  b. J. T. Patton, U.S. Pat. 3,101,374 (1963).
  c. S. L. Cooper and A. V. Tobolsky, *Text. Res. J.* **36**, 800 (1966).
  d. H. J. Hagenmeyer, Jr. and M. B. Edwards, U.S. Pat. 3,529,037 (1970).
  e. A. Eisenberg, *Macromolecules*, **3**, 147 (1970).

two-phase morphology to simulate the mechanical characteristics of a reinforced, crosslinked elastomer. By being able to be processed at elevated temperatures, the desired shape is easily obtained. On cooling below the T_g of the plastic component, usually polystyrene, a physically crosslinked elastomer is obtained.

The thermoplastic elastomers shown in Table 2.3 come in two types. One is the triblock copolymer type manufactured by the Shell Chemical Co. under the name of Kraton®. Star block copolymers,[32] which contain a number of radial arms of elastomer emanating from a central source, each tipped with a plastic block, are made by the Phillips Petroleum Co. under the name of Solprene®.

Of course, the applications shown above always result from synergistic behavior. So far, there have been many more failures, where the material is no better, and frequently worse, than the arithmetic average of the two homopolymers.

SUGGESTED READING. RECENT MULTICOMPONENT POLYMER BOOKS

S. L. Aggarwal, *Block Polymers*, Plenum, New York (1970).

D. C. Allport and W. H. Janes, eds., *Block Copolymers*, Applied Science, Barking, England (1973).

C. B. Bucknall, *Toughened Plastics*, Applied Science, London (1977).

J. J. Burke and V. Weiss, eds., *Block and Graft Copolymers* Syracuse University, Syracuse, New York (1973).

A. Casale and R. S. Porter, *Polymer Stress Reactions*, Vol. 1, *Introduction*, Academic, New York (1978).

A. Casale and R. S. Porter, *Polymer Stress Reactions*, Vol. 2, *Experiments*, Academic, New York (1979).

R. J. Ceresa, *Block and Graft Copolymerization*, Vol. 1, Wiley, New York (1973).

S. L. Cooper and G. M. Estes, eds., *Multiphase Polymers*, Advances in Chemistry Series No. 176, American Chemical Society, Washington D.C. (1979).

A. Eisenberg and M. King, *Ion Containing Polymers*, Academic, New York (1977).

H. S. Katz and J. V. Milewski, *Handbook of Fillers and Reinforcements for Plastics*, Van Nostrand Reinhold, New York (1978).

H. H. Kausch, *Polymer Fracture*, Springer, New York (1978).

D. Klempner and K. C. Frisch, eds., *Polymer Alloys—Blends, Blocks, Grafts and Interpenetrating Networks*, Polymer Science and Technology Vol. 10, Plenum, New York (1977).

D. Klempner and K. C. Frisch, eds., *Polymer Alloys II*, Plenum, New York (1980).

Yu. S. Lipatov and L. M. Sergeeva, *Interpenetrating Polymeric Networks*, Naukova Dumka, Kiev (1979).

J. A. Manson and L. H. Sperling, *Polymer Blends and Composites*, Plenum, New York (1976).

L. E. Nielsen, *Mechanical Properties of Polymers and Composites*, Vols I and II, Marcel Dekker, New York (1974).

L. E. Nielsen, *Predicting the Properties of Mixtures—Mixture Rules in Science and Engineering*, Marcel Dekker, New York (1978).

A. Noshay and J. E. McGrath, *Block Copolymers—Overview and Critical Survey*, Academic, New York (1977).

O. Olabisi, L. M. Robeson, and M. T. Shaw, *Polymer–Polymer Miscibility*, Academic, New York (1979).

D. R. Paul and S. Newman, eds., *Polymer Blends*, Vols. 1 and 2, Academic, New York (1978).

N. A. J. Platzer, *Copolymers, Polyblends, and Composites*, Advances in Chemistry Series No. 142, American Chemical Society, Washington, D.C. (1975).

N. A. J. Platzer, ed., *Multicomponent Polymer Systems*, Advances in Chem. Series No. 99, American Chemical Society, Washington, D.C. (1971).

L. H. Sperling, ed., *Recent Advances in Polymer Blends, Grafts, and Blocks*, Vol. 4, Plenum, New York (1974).

REFERENCES

1. H. Tompa, *Polymer Solutions*, Butterworth, London (1956).
2. R. L. Scott, *J. Chem. Phys.* **17**, 279 (1949).
3. S. Krause, in *Colloidal and Morphological Behavior of Block and Graft Copolymers*, G. E. Molau, ed., Plenum, New York (1971).
4. T. Inoue, T. Soen, T. Hashimoto, and H. Kawai, in *Block Copolymers*, S. L. Aggarwal, ed., Plenum, New York (1970).
5. D. J. Meier, *J. Polym. Sci.* **26C**, 81 (1969).
6. A. A. Donatelli, L. H. Sperling, and D. A. Thomas, *J. Appl. Polym. Sci.* **21**, 1189 (1977).
7. P. J. Flory, *Principles of Polymer Chemistry*, Cornell University, Ithaca, New York (1953).
8. C. B. Bucknall, *Toughened Plastics*, Applied Science, London (1977).
9. C. B. Bucknall and T. Yoshi, *Br. Polym. J.*, **10**, 53 (1978).
10. E. Helfand and Z. R. Wasserman, *Polym. Eng. Sci.* **17**, 582 (1977).
11. E. Helfand, *Macromolecules* **8**, 552 (1975).
12. E. Helfand and A. M. Sapse, *J. Chem. Phys.* **62**, 1327 (1975).
13. D. J. Meier, in *The Solid State of Polymers*, P. H. Geil, E. Baer, and H. Wada, eds., Marcel Decker, New York (1974).
14. L. P. McMaster, *Macromolecules* **6**, 760 (1973).
15. L. P. McMaster, in *Copolymers, Polyblends, and Composites*, N. A. J. Platzer, ed., Advances in Chemistry Series No. 142, American Chemical Society, Washington, D.C. (1975).
16. P. J. Flory, R. A. Orowoll, and A. Vrij, *J. Am. Chem. Soc.* **86**, 3515 (1964).
17. P. J. Flory, *J. Am. Chem. Soc.* **87**, 1833 (1965).
18. D. Patterson and A. Robard, *Macromolecules* **11**, 691 (1978).
19. I. C. Sanchez, in *Polymer Blends*, Vol. I, , D. R. Paul and S. Newman, eds., Academic, New York (1978).
20. T. K. Kwei and T. T. Wang, in *Polymer Blends*, Vol. I, D. R. Paul and S. Newman, eds. Academic, New York (1978).
21. R. E. Bernstein, C. A. Cruz, D. R. Paul, and J. W. Barlow, *Macromolecules* **10**, 681 (1977).
22. A. Robard and D. Patterson, *Macromolecules* **10**, 1021 (1977).
23. F. E. Karasz and W. J. MacKnight, in *Contemporary Topics in Polymer Science*, Vol. 2, E. M. Pearce and J. R. Schaefgen, eds., Plenum, New York (1977).
24. A. A. Donatelli, L. H. Sperling, and D. A. Thomas, *Macromolecules* **9**, 671, 676 (1976).
25. G. C. Eastmond and E. G. Smith, *Polymer* **17**, 367 (1976).

26. G. C. Eastmond and D. G. Phillips, in *Polymer Alloys*, D. Klempner and K. C. Frisch, eds., Plenum, New York (1977).

27. R. R. Durst, R. M. Griffith, A. J. Urbanic, and W. J. Van Essen, in *Toughness and Brittleness of Plastics*, R. D. Deanin and A. M. Crugnola, eds., Advances in Chemistry Series No. 154, American Chemical Society, Washington, D.C. (1976).

28. S. L. Aggarwal, *Polymer* **17**, 938 (1976).

29. S. Marti and G. Riess, *Makromol. Chem.* **179**, 2569 (1978).

30. S. L. Cooper and G. M. Estes, *Multiphase Polymers*, Advances in Chemistry Series No. 176, American Chemical Society, Washington, D.C. (1979).

31. A. Noshay and J. E. McGrath, *Block Copolymers—Overview and Critical Survey*, Academic, New York (1977).

32. L. Bi and L. J. Fetters, *Macromolecules* **9**, 732 (1976).

33. D. R. Paul and S. Newman, *Polymer Blends*, Vols. I and II, Academic, New York (1978).

34. Yu. S. Lipatov and L. M. Sergeeva, *Interpenetrating Polymeric Networks*, Naukova Dumka, Kiev (1979).

35. L. H. Sperling, *J. Polym. Sci. Polym. Symp.* **60**, 175 (1977).

36. D. A. Thomas, *J. Polym. Sci. Polym. Symp.* **60**, 189 (1977).

37. M. Shen and H. Kawai, *AIChE J.*, **24**(1), 1 (1978).

38. L. H. Sperling, *Polym. Eng. Sci.* **16**, 87 (1976).

39. M. Matsuo, C. Nozaki, and Y. Jyo, *Polym. Eng. Sci.* **9**, 197 (1969).

40. M. Matsuo, *Jpn. Plast.* **2** (July), 6 (1968).

41. J. F. Beecher, L. Marker, R. D. Bradford, and S. L. Aggarwal, *J. Polym. Sci.*, **26C**, 117 (1969).

42. G. Kraus and K. W. Rollman, *J. Polym. Sci. Polym. Phys. Ed.* **14**, 1133 (1976).

A NOMENCLATURE FOR MULTIPOLYMER SYSTEMS

3.1. EXISTING NOMENCLATURE

The previous two chapters briefly described a number of two-polymer combinations: polymer blends, blocks, grafts, and IPNs. A few somewhat more complicated systems were alluded to: blends of a homopolymer with a block copolymer, or a mixture of a graft copolymer with one or both homopolymers. This chapter will explore some of the more complex (and interesting) structures, and provide a nomenclature scheme where one now does not exist.[1]

A good nomenclature scheme should accomplish two tasks: (1) accurately name all of the existing known compositions and (2) provide a present and future base for making systematic information retrieval searches for known, missing, unsynthesized, or unrecognized materials included within the nomenclature scheme.

Many simple polymer grafts and blocks already have precise names.[2–8] For example, poly(butadiene-b-styrene) represents the structure

$$A-A-A-\cdots-A-B-B-B-\cdots-B \qquad (3.1)$$

where A stands for the butadiene mer and B stands for the styrene mer in the block copolymer arrangement, as indicated by the small $-b-$.

Similarly, poly(butadiene-g-styrene) is the name for the graft copolymer

$$\begin{array}{c} A-A-A-A-A-A-\cdots-A \\ | \\ B-B-B-B-\cdots-B \end{array} \qquad (3.2)$$

where it is understood that the symbol $-g-$ means graft copolymer, the first indicated species, butadiene, forms the backbone, and the second species, styrene, the side chain. Similarly, the symbol $-co-$ stands for a random copolymer, $-a-$ for an alternating copolymer, and occasionally the symbol $-cl-$ is used for crosslinked systems.

Materials such as polymer blends, IPNs,[9,10] AB-crosslinked poly-mers,[11] and many other combinations have no existing systematic nomen-clature. The situation is further complicated by the fact that the time sequence of polymerization, grafting, and/or crosslinking each produce materials possessing different morphologies and often widely different mechanical or physical properties.[7]* An examination of the patent lit-erature reveals complex combinations of up to five polymers. See Table 8.1 for some examples. An exact description of each requires a long paragraph, and the identification of the isomeric possibilities presents a serious chal-lenge even to the expert, leading to the need for a more comprehensive nomenclature.

Because of the above considerations, Sperling and co-workers[12,13] have evolved a tentative nomenclature scheme for polymer blends, blocks, graft copolymers, and IPNs. A few final points are in order.

1. From a notation point of view, the system to be presented reads from left to right, corresponding to ordinary chemical notation.[2-4] Some of the earlier papers had notation which read from right to left, following standard mathematical notation. However, since the following is primarily intended for people working in the chemical arts, the system now reads in the sequence ordinarily used by chemists.

2. The nomenclature system has been submitted to the Nomenclature Committee of the Polymer Division of the American Chemical Society[14] for evaluation relative to adoption in full or in part. Thus, at the time of this writing, it has not yet achieved any official status. Rather, it is presented so that the interrelationships among the numerous chemical possibilities will become clearer.†

3.2. INTRODUCTION TO THE NEW SYSTEM

More than 200 distinct topological methods of multipolymer organiza-tion are already known[7,12] and new topologies are being reported frequently. In addition to the usual molecular specifications, the time sequence of synthesis is important in many cases and needs to be preserved in the nomenclature for a full comprehension of the final product.

The following nomenclature scheme uses a short list of elements (Table 3.1), polymers and polymer-reaction products, which are reacted together in

---

* The introduction of the time sequence of reaction clearly steps out of the bounds of classical chemistry. However, spatial and morphological aspects do lead to different products on the practical level.

† Section 3.2 describes the general nomenclature using the concepts of mathematical rings. The proposal to the Nomenclature Committee emphasizes the notation in Section 3.3 (see eqn. 3.25) and Appendix 3.2.

Table 3.1. Basic Elements[8]

Symbol	Meaning
P	Linear polymer
R	Random copolymer
A	Alternating copolymer
M	Mechanical or physical blend
G	Graft copolymer
C	Crosslinked polymer
B	Block copolymer
S	Starblock copolymer
I	Interpenetrating polymer network
U	Unknown reaction mixtures

specific ways by binary operations (a joining of two elements) (Table 3.2). A series of numerical subscripts are used on the elements to allow an arbitrarily large number of different polymers to be designated conveniently.

This nomenclature primarily answers three questions about the chemical entity: (1) which polymers are combined, (2) principal modes of combination, and (3) the time sequence of reaction. Other items of information, such as weight proportions, molecular weights, morphology, tacticity, etc. can be included as ancillary items but will not be discussed in detail below.

While the nomenclature scheme was originally derived to describe polymer blends, blocks, grafts, and IPNs, generalizing the scheme to include random and alternating copolymers and starblock copolymers made the

Table 3.2. Binary Operations and Associated Reactions[8]

Symbol	Reaction induced
O_R	Random copolymer formation
O_A	Alternating copolymer formation
O_M	Mechanical or physical blending[a]
O_G	Graft copolymer formation
O_C	Crosslinked network formation
O_B	Block copolymer formation
O_S	Starblock copolymer formation
O_I	Interpenetrating polymer network
O_U	Unknown reaction mixtures

[a] A blend indicates a mixture of two or more molecular species without chemical bonding between them and can be induced by mechanical blending, physical means such as coprecipitation, or chemical means such as degrafting.

scheme more useful. Although the random and alternating copolymers are formed through direct combination of monomers, rather than polymers, their importance warrants their inclusion. Through the addition of further operations, additional types of structures can easily be included.

As with other nomenclature schemes, the objective will be to name the most important product(s) of a reaction. In some cases, several isomers are possible, but listing each one all the time becomes cumbersome and perhaps misleading.

Subscripts. The subscripts, $1, 2, 3, \ldots, i, j, \ldots$ will indicate a numbering of the polymers:

$$P_1, \quad \text{polymer 1} \tag{3.3}$$

$$Pj, \quad \text{polymer } j \tag{3.4}$$

When more than one subscript appears, the first appearing subscript indicates the first formed polymer, and the second subscript, the second formed polymer, as shown in Table 3.3. In R, A, and M materials, as defined in Table 3.1, the time sequence is unimportant. The first listed subscript should describe the most important component.

3.2.1. Reaction Examples

The binary operation symbol used between any two elements indicates the required manner of their reaction together. (The reader is reminded that all combinations, mathematical or chemical, are binary operations. The advent of modern computers has focused much attention on this often neglected fact.) The first formed element appears on the left and the reaction-time sequence (when required) proceeds from left to right.

Table 3.3. Illustration of the Use of Subscripts[8]

P_1	A homopolymer composed of monomer 1 mers.
G_{12}	A graft copolymer having polymer 1 as the backbone and polymer 2 as the side chain.
C_{11}	A crosslinked polymer composed of a network of polymer 1 chains.
R_{27}	A random copolymer composed of monomers 2 and 7. (In R and A materials, the time element can be replaced by the composition importance order.)
U_{12}	Unknown reaction mixture of polymer 1 and polymer 2. Specific examples: mechanochemical blends which contain unknown proportions of grafts, blocks, and homopolymers or solution graft copolymers which contain much homopolymer. Various isomers can be formed.
M_{21}	Mechanical or chemical blend, with no time sequence. First subscript listed may have the higher weight percentage, etc.

The following examples illustrate the use of this notation. A graft copolymer having polymer I* as the backbone and polymer II as the side chain, synthesized in that order, is represented by

$$P_1 \, O_G \, P_2 = G_{12} \tag{3.5}$$

a crosslinked copolymer made from polymer i and j, by

$$P_i \, O_C \, P_j = C_{ij} \tag{3.6}$$

(an AB-crosslinked copolymer). An alternating copolymer of mers III and V is represented by

$$P_3 \, O_A \, P_5 = A_{35} \tag{3.7}$$

(3.8) represents a physical blend of graft copolymers G_{13} and G_{23}, where polymers I and II were blended first, followed by the polymerization of polymer III with grafting onto polymers I and II:

$$(P_1 \, O_M \, P_2) \, O_G \, P_3 = G_{13} \, O_M \, G_{23} \tag{3.8}$$

On the other hand, in (3.9), the symbol $G_{1(2,3)}$ indicates both polymers grafted onto polymer I:

$$P_1 \, O_G \, (P_2 \, O_M \, P_3) = G_{12} \, O_M \, G_{13} \, O_M \, G_{1(2,3)} \tag{3.9}$$

[Alternately, the symbol $U_{1(2,3)}$ might be used since the exact modes of grafting cannot be specified, or which product(s) will dominate.] The structures previously described in Figure 1.3 may now be described according to the following new system:

$$\text{(a)} \qquad M_{12} \tag{3.10}$$

$$\text{(b)} \qquad G_{12} \tag{3.11}$$

$$\text{(c)} \qquad B_{212} \tag{3.12}$$

$$\text{(d)} \qquad C_{11} \, O_I \, P_2 \tag{3.13}$$

$$\text{(e)} \qquad C_{11} \, O_I \, C_{22} \tag{3.14}$$

$$\text{(f)} \qquad C_{12} \tag{3.15}$$

The above examples use the abstract element symbols P_1, G_{12}, etc. Real monomers, polymers, and multipolymer combinations can be substituted using chemical notation instead of the elements but retaining the binary operation notation.

---

* In keeping with other IPN-related literature, Roman numerals will indicate the polymer in the text, and Arabic numerals will be used in the subscripts.

More than one notation appears possible.

1. *Direction Substitution of Element Notation.* The following represents the grafting of polystyrene side chains onto a polybutadiene backbone:

$$\text{polybutadiene } O_G \text{ polystyrene} \tag{3.16}$$

An AB-crosslinked copolymer of Bamford and Eastmond[15] is

$$\text{poly(vinyl trichloroacetate) } O_C \text{ polystyrene} \tag{3.17}$$

Here, two polymers are caused to form one network through conterminous junctions of the ends of the polystyrene to the poly(vinyl trichloroacetate).

An IPN of two different polymers is

$$\text{poly(ethyl acrylate) } O_C \text{ poly(ethyl acrylate) } O_I \text{ polystyrene } O_C \text{ polystyrene} \tag{3.18}$$

where the poly(ethyl acrylate) network is formed first, then swollen with styrene and crosslinker, and polymerized *in situ.*[16]

Mechanical blends, usually not considered as single chemical entities, can easily be represented by[17]

$$\text{polyethylene } O_M \text{ polypropylene} \tag{3.19}$$

2. *Retention of Abstract Elemental Symbolism, but Equating Elements and Polymers Separately.* This notation method is more suitable for more complex reaction mixtures. The equivalent of (3.10) can be written as

$$G_{12} \qquad \begin{aligned} P_1 &= \text{polybutadiene} \\ P_2 &= \text{polystyrene} \end{aligned} \tag{3.20}$$

Appendixes 3.1 and 3.2 give further examples and show how the new system can, in significant measure, be related to the older system. The system can also utilize vertical notation modes, as described by Sperling and Corwin[8] in more complex multipolymer combinations.

As one example of vertical notation, the triblock copolymer of polystyrene, polybutadiene, and polystyrene is written in that order. The polybutadiene in this case has been grafted with poly(methyl methacrylate). The proposed nomenclature is shown by

$$
\begin{array}{ll}
P_1 & \\
O_B & P_1 = \text{polystyrene} \\
& P_2 = \text{polybutadiene} \\
P_2\, O_G\, P_3 & P_3 = \text{poly(methyl methacrylate)} \qquad (3.21) \\
O_B & \\
P_1 &
\end{array}
$$

For emphasis, the notation is read from top to bottom, and from left to right.

3.2.2. Simultaneous Mixing or Reacting

One of the advantages of the proposed nomenclature is the preservation of the time sequence of polymer reaction. Examples include grafting reactions and formation of IPNs. In some cases, such as the mechanical blends or random copolymers, the time sequence has no meaning.

In other cases it may be important to indicate that the mixing or reactions were carried out simultaneously. This will be indicated, where necessary, by brackets, e.g.,

$$[C_{11}\, O_I\, C_{22}] \qquad (3.22)$$

This indicates a simultaneous interpenetrating network, where C_{11} was formed by one reaction (say, addition), C_{22} was formed independently by another reaction (say, condensation), and the two networks are simultaneously polymerized. Without brackets a sequential synthesis of an interpenetrating polymer network is indicated:

$$C_{11}\, O_I\, C_{22} \qquad (3.23)$$

In both cases, this special mode of interpenetration is indicated by the binary operation O_I.

3.2.3. Coefficients and Molecular Weights

Where proportions are known, coefficients can be used:

$$P_1\, O_R\, 5P_2 \qquad (3.24)$$

This indicates 5 mol of P_2 per mole of P_1. Molecular weights, where known, may be indicated by parentheses following the element: $P_1(5 \times 10^4\,\text{g/mol})\, O_G\, P_2(3 \times 10^5\,\text{g/mol})$.

3.3. NOMENCLATURE IN THIS WORK

Appendix 3.2 describes the use of -*i*- in the polymer name to indicate an interpenetrating polymer network. Thus an IPN of poly(ethyl acrylate) and polystyrene, polymerized in that order, would be written

$$\text{poly(ethyl acrylate-}i\text{-styrene)} \qquad (3.25)$$

This is a notation proposed to the Nomenclature Committee of the American Chemical Society's Polymer Division,[1,8,18] but not yet formally accepted or in use anywhere. Most of the IPN literature until now has used a slash, albeit also totally without formal recognition. Written with a slash, (3.25) becomes

$$\text{poly(ethyl acrylate)/polystyrene} \qquad (3.26)$$

It must be emphasized again that the first polymerized network appears first, and the second network is written second.

In order to maintain greater continuity with the prior literature, the slash notation will be used in this work. In a number of places, the notation described in (3.5)–(3.15) will be employed, as it provides a concise statement of the structure under discussion. However, the text is written in such a way as to be understood without employing the proposed notation.

While both (3.25) and (3.26) carry the most important part of the information, i.e., if the structure is an IPN and not, say, a graft copolymer, some information is lost. Thus, in addition to the notation described in this chapter, sequential IPNs will be differentiated from simultaneously polymerized materials, SINs, and information relating to weight ratios, crosslink densities, etc. will be included for greater clarity.

3.4. ABBREVIATIONS AND JARGON

All scientific and technical fields have a plethora of abbreviations and jargon. Some of the more important terms employed in the field of interpenetrating polymer networks include the following.

1. IPN: interpenetrating polymer network, the general term. Also used to indicate the time-sequential synthesis product.

2. IEN: interpenetrating elastomeric network, originally used by Frisch *et al.*[19] to denote materials made by mixing two latexes, coagulating them, followed by independent crosslinking reactions.

3. SIN: simultaneous interpenetrating network, where both polymers are synthesized simultaneously, that is, by chain and step polymerization reactions.

4. Semi-IPN: compositions, generally of one crosslinked polymer and one linear polymer.

5. Semi-I, semi-II: semi-IPNs where, respectively, polymer I or polymer II is the crosslinked component. Semi-I IPNs were designated "snake cages" in the older literature.

6. Interstitial composites: notation of Allen et al.[20] for materials which would be described as semi-SINs in the above notation system; both polymers are synthesized simultaneously, but one is linear and the other is crosslinked.

7. Gradient IPN: an IPN of nonuniform macroscopic composition, usually by nonequilibrium swelling in monomer II, and polymerizing rapidly.

8. Homo-IPN: an IPN where both polymers are identical. This term replaces the earlier notation "Millar IPNs" because it is more descriptive of the materials.

Some additional abbreviations include the following:

1. ABCP: AB-crosslinked polymer
2. HiPS: high-impact polystyrene
3. ABS: acrylonitrile–butadiene–styrene. An improved rubber-toughened polystyrene
4. ABA: order of blocks in a triblock copolymer

Many common polymers are referred to by an abbreviation:

 1. PS: polystyrene
 2. PB: polybutadiene
 3. PEA: poly(ethyl acrylate)
 4. PMMA: poly(methyl methacrylate)
 5. PU: polyurethane (generic term)
 6. PVC: poly(vinyl chloride)
 7. SBR: styrene–butadiene rubber, random copolymer
 8. PDMS: poly(dimethyl siloxane)
 9. NBR: nitrile rubber; poly(butadiene-co-acrylonitrile)
10. TEGDM: tetraethylene glycol dimethacrylate
11. TMPTM: trimethylolpropane trimethacrylate
12. SBS: styrene–butadiene–styrene triblock copolymer
13. SEBS: styrene–ethylene-co-butylene–styrene triblock copolymer

APPENDIX 3.1. EXAMPLES OF THE PROPOSED NOMENCLATURE

In the following the several systems of nomenclature, where they exist, will be compared to the proposed system. A series of examples of increasing complexity will be given and each of the proposed elements and/or binary operations briefly discussed.

1. *Homopolymers.* Homopolymers are designated by the element P. When used alone, subscript numbers are not required, but the subscripting yields greater clarity for the general case:

$$\text{a.} \quad \text{cellulose} = P_1 \qquad\qquad (3\text{A}.1)$$

$$\text{b.} \quad \text{polystyrene} = P_4 \qquad\qquad (3\text{A}.2)$$

2. *Alternating Copolymers.* These materials are designated by the element A and require two numerical subscripts, one for each mer.

a. Poly(styrene-*alt*-maleic acid anhydride): This reaction is designated by

$$P_1 O_A P_2 = A_{12} \qquad P_1 = \text{polystyrene}$$
$$P_2 = \text{poly(maleic acid anhydride)} \qquad (3\text{A}.3)$$

The element A_{12} specifies the final product.

3. *Random Copolymers*

a. Poly(vinyl acetate-*co*-butadiene): This polymer is designated by the element R. For the above composition,

$$R_{34} \qquad P_3 = \text{poly(vinyl acetate)}$$
$$P_4 = \text{polybutadiene} \qquad (3\text{A}.4)$$

Note: Homopolymers, alternating polymers, and random copolymers are adequately named by the process-based nomenclature already in existence. The proposed nomenclature is given to show the generality of the system, for completeness.

4. *Block Polymers.* A diblock polymer has the general structure

$$P_1 \qquad P_2 \qquad\qquad (3\text{A}.5)$$

where the chains are joined end-to-end.

a. Poly(styrene-*b*-butadiene): This polymer combination is designated by

$$B_{12} \qquad P_1 = \text{polystyrene}$$
$$P_2 = \text{polybutadiene} \qquad (3\text{A}.6)$$

b. Poly(styrene-*b*-butadiene-*b*-styrene):

$$B_{121} \qquad P_1 = \text{polystyrene}$$
$$P_2 = \text{polybutadiene} \qquad (3A.7)$$

c. A multiblock polymer can be written by the appropriate series of subscripts:

$$B_{12321} \qquad P_1 = \text{polystyrene}$$
$$P_2 = \text{polybutadiene} \qquad (3A.8)$$
$$P_3 = \text{poly(vinyl chloride)}$$

In the above, a block polymer is composed of a set of homopolymers. If one of them is a random copolymer, the term block copolymer should be used.

d. Poly[vinyl chloride-*b*-(styrene-*co*-butadiene)] can be designated by

$$P_1 O_B R_{23} \qquad P_1 = \text{poly(vinyl chloride)}$$
$$P_2 = \text{polystyrene} \qquad (3A.9)$$
$$P_3 = \text{polybutadiene}$$

5. *Graft Polymers.* These compositions have the second polymer attached to the backbone of the first; the exact position is random. This may be pictured

$$(3A.10)$$

a. Poly(butadiene-*g*-styrene):

$$G_{12} \qquad P_1 = \text{polybutadiene}$$
$$P_2 = \text{polystyrene} \qquad (3A.11)$$

b. Now, somewhat more complex structures can be introduced. For example, polybutadiene grafted with both polystyrene and (then) poly(vinyl acetate) cannot be easily designated with the process-oriented system:

$$\text{poly[(butadiene-}g\text{-styrene)-}g\text{-vinyl acetate]} \qquad (3A.12)$$

for example, does not specify where the vinyl acetate is situated. The proposed nomenclature would treat the compositions as follows:

$$P_1 \, O_G \, P_3 \qquad P_1 = \text{polybutadiene}$$
$$O_G \qquad\qquad P_2 = \text{polystyrene} \qquad\qquad (3A.13)$$
$$P_2 \qquad\qquad P_3 = \text{poly(vinyl acetate)}$$

indicates the structure

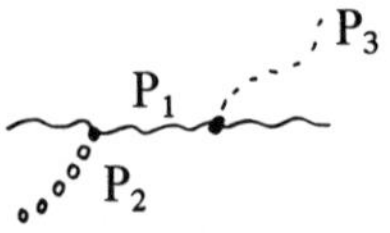

$$(3A.14)$$

while

$$P_1 \, O_G \, P_2 \, O_G \, P_3 = G_{123} \qquad\qquad (3A.15)$$

indicates the structure

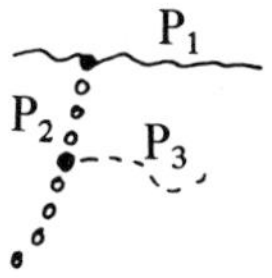

$$(3A.16)$$

6. *Starblock Polymers.* While many star-shaped structures can be formed, the usual structure referred to may be designated

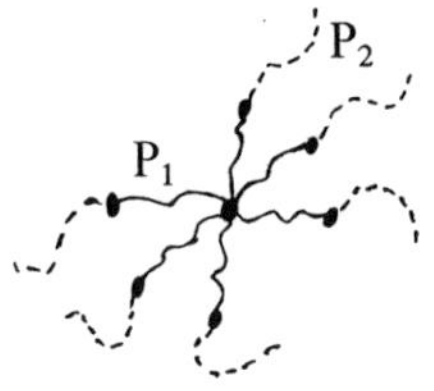

$$(3A.17)$$

where the star polymer of P_1 is reacted in block polymer form by P_2. No accepted nomenclature exists here. The proposed nomenclature would treat the above as follows:

$$P_1 \, O_S \, P_2 = S_{12} \qquad\qquad (3A.18)$$

If the number (or average number) of arms is known, this information could follow in parentheses:

$$S_{12} \, (6) \qquad P_1 = \text{polybutadiene}$$
$$\qquad\qquad\qquad\qquad\qquad\qquad\qquad\qquad (3A.19)$$
$$P_2 = \text{polystyrene}$$

which indicates a starblock polymer having six arms, the inner block of each arm being polybutadiene and the outer block being polystyrene.

The next several multipolymer combinations have no accepted standard nomenclature.

7. *Mechanical or Physical Blends.* While a blend is a physical mixture rather than a definite reacted species, many block and graft polymers contain greater or lesser quantities of homopolymer. In particular, the only slightly grafted HiPS and ABS materials are usually listed together with the polybutadiene/polystyrene mechanical blends. It is convenient to introduce a special designation for blended materials: A simple example is

$$M_{12} \qquad P_1 = \text{starch}$$
$$P_2 = \text{natural rubber} \qquad (3A.20)$$

8. *Branched and Crosslinked Polymers.* A branched homopolymer may be written

$$G_{11} \qquad P_1 = \text{starch} \qquad (3A.21)$$

i.e., a graft of a polymer onto its own species. A crosslinked homopolymer is designated by the element C. The structure

$$(3A.22)$$

symbolizes the crosslinking of one polymer with itself. As an example,

$$P_1 \, O_C \, P_1 = C_{11} \qquad P_1 = \text{natural rubber} \qquad (3A.23)$$

9. *AB-Crosslinked Polymers.* Sometimes abbreviated ABCPs indicate two polymers grafted together to form one continuous network. The general structure can be idealized:

$$(3A.24)$$

Note that P_2 is grafted on both ends in a conterminous fashion. The element G_{12} or the use of the process-oriented -*g*- symbolism fall short. Noting that this is a crosslinked network of two polymers, the notation

$$C_{12} \qquad P_1 = \text{poly(vinyl trichloroacetate)}$$
$$P_2 = \text{polystyrene} \qquad (3A.25)$$

provides a simple designation. (Note that C_{11} represents a crosslinked homopolymer of P_1.)

10. *Interpenetrating Polymer Networks*, abbreviated IPNs. No standard nomenclature exists. As an interpenetrating blend of two crosslinked polymers, the structure can be represented as

$$(3A.26)$$

While the name poly(ethyl acrylate-*i*-styrene) might serve, the designation

$$C_{11} O_1 C_{22} = I_{12} \qquad (3A.27)$$

yields greater exactness. For example, if a more complex material is called for, e.g., one of the networks is an ABCP, the notation

$$I_{12,33} \qquad P_1 = \text{poly(vinyl trichloroacetate)}$$

$$P_2 = \text{polystyrene} \qquad (3A.28)$$

$$P_3 = \text{poly(vinyl acetate)}$$

serves more exactly.

Note that with ABCPs, IPNs, and many other examples cited above, the time order of polymerization is important; inverting the time order of reactions may yield distinguishable products.

11. *Unknown Reaction Mixtures.* It is convenient to introduce the symbol U to indicate an unknown mixture of species. This is important for complex materials and also for novel reactions in general.

12. *Analysis of a Literature Composition.* B. Vollmert, in Example No. 7 of the U.S. Patent 3,055,859 (1962) provides an interesting example. He wrote:

Example 7

I. An emulsion consisting of 2,000 parts of water, 10 parts of an alkyl (C_{12} to C_{14}) sulfonate, 960 parts of butyl acrylate, 40 parts of acrylic acid, 2.5 parts of 1.4 butanediol diacrylate, 1 part of potassium persulfate and 2.5 parts of dodecyl mercaptan is polymerized under nitrogen for 5 hrs at 60° to 65°C. (K-value according to Fikentscher in benzene about 50.)

II. An emulsion consisting of 2,000 parts of water, 10 parts of an alkyl (C_{12} to C_{14}) sulfonate, 530 parts of butyl acrylate, 430 parts of styrene, 40 parts of 1.4-butane-diol monoacrylate and 1 part of potassium persulfate is polymerized under nitrogen for 5 hrs at 60°C. (K-value according to Fikentscher in benzene about 150.)

III. An emulsion consisting of 2,000 parts of water, 10 parts of an alkyl (C_{12} to C_{14}) sulfonate, 994 parts of styrene, 6 parts of 1.4-butane-diol monoacrylate, 1 part of potassium persulfate and 1 part of normal dodecyl mercaptan is polymerized under nitrogen at a pH of 8.5 for 6 hrs at 65°C. (K-value according to Fikentscher in benzene about 70.)

IV. 994 parts of styrene are polymerized with 6 parts of 1.4–butane-diol monoacrylate and 1 part of azo-isobutyronitrile under nitrogen first for 24 hrs at 60°C, then for 24 hrs at 120°C. (K-value according to Fikentscher about 70.)

150 parts of emulsion I, 150 parts of emulsion II and 200 parts of emulsion III are mixed and dried on a spray roller drier. 500 parts of the resultant granular powder are mixed with 500 parts of the polymer IV in an endless screw. The granulate thus prepared is kept for 24 hrs at 180°C. under nitrogen free from oxygen and treated for 10 min on a roller running with friction at 140°C.

A homogeneous white translucent plastic composition is obtained which can be worked up by injection molding or vacuum deep drawing to shaped articles having high impact strength and good thermal stability.

Under the assumption that the carboxyl groups in polymer 1 react with the free hydroxyl groups in polymers 2, 3, and 4 to form a polyester graft site, the composition can be named

$$C_{11} \, O_G \, (P_2 \, O_M \, P_3 \, O_M \, P_4)$$

$P_1 = $ poly(butyl acrylate-co-acrylic acid)

$P_2 = $ poly(butyl acrylate-co-styrene)

$P_3 = $ polystyrene (emulsion)   (3A.29)

$P_4 = $ polystyrene (bulk)

APPENDIX 3.2. RELATING THE NEW SYSTEM TO THE OLD

To the greatest extent possible, of course, continuity of ideas must be maintained. Therefore the examples appearing in Appendix 3.1 will be recast into a somewhat more standard form. The following symbols will be employed:

-co-	random copolymer
-alt-	alternating copolymer
-b-	block copolymer junction
-g-	graft copolymer junction (also employed for branched homopolymers)
-c-	crosslink junction (also for conterminous graft networks)
-s-	starblock copolymer junction
-m-	mechanical blend
-i-	interpenetrating polymer network.

In the following, the numbers in parentheses on the left refer to equation numbers in Appendix 3.1. An asterisk beside the equation number indicates that the new nomenclature is identical with the standard nomenclature.

(2)* polystyrene
(3)* poly(styrene-*alt*-maleic acid anhydride)
(4)* poly(vinyl acetate-*co*-butadiene)
(6)* poly(styrene-*b*-butadiene)
(10)* poly[vinyl chloride-*b*-(styrene-*co*-butadiene)]

(13) $\text{poly}\left[\text{butadiene-}\begin{bmatrix} g\text{-styrene} \\ g\text{-vinyl acetate} \end{bmatrix}\right]$

(15) poly(butadiene-*g*-styrene-*g*-vinyl acetate)
(19) poly(butadiene-*s*-styrene)
(20) poly(starch-*m*-natural rubber)
(21) starch-*g*-starch
(23) natural rubber-*c*-natural rubber
(24) poly(vinyl trichloroacetate-*c*-styrene)
(27) poly(ethyl acrylate-*i*-styrene)
(28) poly [(vinyl trichloroacetate-*c*-styrene)-*i*-vinyl acetate]
(29) poly [(butyl acrylate-*co*-acrylic acid)-*c*-

$\text{(butyl acrylate-}co\text{-acrylic acid)-}\begin{bmatrix} \text{-}g\text{-(butyl acrylate-}co\text{-styrene)} \\ \text{-}g\text{-styrene(emulsion)} \\ \text{-}g\text{-styrene(bulk)} \end{bmatrix}$

In equations (13) and (29) the time order of reaction is preserved through the use of consecutive vertical components. In other cases, such as equations (24), (27), and (28), the time sequence reads from left to right.

REFERENCES

1. L. G. Donaruma, *J. Chem. Inf. Comput. Sci.* **19**, 68 (1979).
2. "IUPAC Nomenclature of Regular Single-Stranded Organic Polymers," *Pure Appl. Chem.* **48**, 373 (1976).
3. *Basic Definitions of Terms Relating to Polymers*, pp. 479–491, IUPAC, Butterworths, London (1974).
4. *List of Standard Abbreviations (Symbols) for Synthetic Polymers and Polymer Materials*, pp. 475–576, IUPAC, Butterworths, London (1962).
5. R. J. Ceresa, *Block and Graft Copolymers*, Butterworths, London (1962).
6. W. J. Burlant and A. S. Hoffman, *Block and Graft Polymers*, Reinhold, New York (1960).
7. J. A. Manson and L. H. Sperling, *Polymer Blends and Composites*, Plenum, New York (1976).
8. L. H. Sperling and E. M. Corwin, in *Multiphase Polymers*, S. L. Cooper and G. M. Estes, eds., Advances in Chemistry Series No. 176, American Chemical Society, Washington, D.C. (1979).

9. E. A. Neubauer, D. A. Thomas and L. H. Sperling, *Polymer* **19**, 188 (1978).

10. D. Klempner and K. C. Frisch, *J. Elastoplast.* **5**, 196 (1973).

11. G. C. Eastmond and D. G. Phillips, in *Polymer Alloys*, D. Klempner and K. C. Frisch, eds., Plenum, New York (1977).

12. L. H. Sperling and K. B. Ferguson, *Macromolecules* **8**, 691 (1975).

13. L. H. Sperling, K. B. Ferguson, J. A. Manson, E. M. Corwin, and D. L. Siegfried, *Macromolecules* **9**, 743 (1976).

14. G. Donaruma, Vice President for Academic Affairs, New Mexico Institute of Mining & Technology, Socorro, New Mexico, Chairman.

15. C. H. Bamford and G. C. Eastmond, *Recent Advances in Polymer Blends, Grafts, and Blocks*, L. H. Sperling, ed., Plenum, New York (1974).

16. V. Huelck, D. A. Thomas and L. H. Sperling, *Macromolecules* **5**, 340, 348 (1972).

17. R. D. Deanin and M. F. Sansone, *Polym. Prepr. Am. Chem. Soc. Div. Polym. Chem.* **19**(1), 211 (1978).

18. L. H. Sperling, to be published.

19. H. L. Frisch, D. Klempner, and K. C. Frisch, *J. Polym. Sci.* **B7**, 775 (1969).

20. G. Allen, M. J. Bowden, D. J. Blundell, F. G. Hutchinson, G. M. Jeffs, and V. Vyvoda, *Polymer* **14**, 597 (1973).

4

HOMO-IPNs AS MODEL NETWORKS

4.1. POLYSTYRENE/POLYSTYRENE IPNs

Most areas of modern research contain two concurrent factors: a basic aspect, which somehow aims at an improvement in scientific understanding, and an applied aspect, which is directed toward a practical goal such as a new or improved material or process. While one or the other factor frequently predominates in a given piece of work, sometimes the two are inextricable.

This chapter emphasizes the role of IPNs as model polymer compositions. It is concerned with aspects of rubber elasticity, physical and chemical crosslinks, and swelling behavior. While much of Chapters 7 and 8 to follow are much more practically oriented, a study of the molecular structure of IPNs will make more apparent the need for fundamental understanding in arriving at the best possible practical materials.

In recent years, important advances in the theory of rubber elasticity have been made. These include the introduction of the so-called phantom networks by Flory[1] and a two-network model for crosslinks and trapped entanglements by Ferry and co-workers.[2,3] The latter builds on work by Flory[4] and others on networks crosslinked twice, once in the relaxed state, and then again in the strained state. In other studies, Kramer[5] and Graessley[6] distinguished among the three kinds of physical entanglements as crosslink sites: the Bueche–Mullins trap, the Ferry trap, and the Langley trap.

These several papers discuss the contribution, or lack of contribution, of physical crosslinks to retractive stresses in the theory of rubber elasticity.[7] In most sequential IPNs described in this monograph, a crosslinked polymer I is swollen with monomer II, plus crosslinking and activating agents, and monomer II is polymerized *in situ*.[8-13] Chapter 5 is devoted to a review of synthetic detail. If the two polymers are identical, the product will be designated as a homo-IPN. These homo-IPNs, because of reduced (or absent) domain formation, provide model networks suitable for the study of the presence or absence of physical crosslinks, as well as other factors, such as the suggested domination of one network over the other.

49

When both polymer networks are chemically identical, the naive viewpoint suggests that a mutual solution will result with few ways to distinguish network I from network II. As will be shown below, this is emphatically not the case.

There have been four major research reports using IPNs where both networks are identical[14–17] as well as other studies and applications.[18–20]* All four major studies, the basis for the analysis in this chapter, interestingly enough used networks of polystyrene (PS) crosslinked with divinyl benzene (DVB). The first publication on PS/PS-type IPNs was a study of swelling by Millar which appeared in 1960.[15] (IPNs having both networks identical in chemical composition have sometimes been called Millar IPNs, after his pioneering work[14]). Shibayama and Suzuki published a paper in 1966[16] on the modulus and swelling properties of PS/PS IPNs, followed by Siegfried, Manson, and Sperling,[14] who also examined viscoelastic behavior and morphology. Most recently, Thiele and Cohen[17] studied swelling and modulus behavior, and derived a key equation with which to study the swelling behavior of IPNs.

In important aspects, this chapter will describe a reexamination of the homo-IPN data by Siegfried et al.[21,22] The results will be scrutinized in the light of the new theoretical developments.[1–6] For clarity, all PS/PS IPNs will be denoted by two pairs of numbers: vol% DVB in network I/vol% DVB in network II, % network I/% network II.

While all four investigators employed PS/PS IPNs crosslinked with DVB, each differed in important details. Table 4.1 summarizes the principal synthetic variations.[14–17] The extent of swelling and polymerization conditions probably influences the final results.

---

* A word must be interjected about these other homo-IPNs. Lipatov et al.[20] studied epoxy/epoxy homo-IPNs, and found that the T_g of the product declined steadily through four successive swelling and polymerization steps. One might speculate if this was due to an increased number of gauche states in the stretched chains. Siegfried et al.[14] found greater relaxation effects than expected, but interpreted the data in terms of the low crosslinking level in polymer I.

Clark[18] employed poly(dimethyl siloxane) (PDMS) homo-IPNs to make improved adhesives. Three separate linear PDMS chains were mixed, each with reactive groups. Polymers I and II reacted to form a network, yielding a semi-IPN. The remaining linear polymer provided the adhesive properties. After adhering the two required surfaces together, raising the temperature initiated a self-crosslinking of polymer III to form the IPN.

Years earlier, Staudinger and Hutchinson[20] employed a homo-IPN of acrylic composition to make optically smooth surfaces. As amplified in Chapter 1, network I was swollen with more monomer of the same type to smooth out surface wrinkles by the stretching incurred on swelling. Polymerization of the new monomer yielded a homo-IPN.

Table 4.1. Synthetic Details for DVB/DVB PS/PS IPNs[22]

Investigator	Methods	Comments
Millar[15]	Suspension	Each sample swelled to equilibrium.
Shibayama and Suzuki[16]	Bulk	Each sample swelled to equilibrium.
Siegfried, Manson and Sperling[14]	Bulk	Controlled degrees of swelling.
Thiele and Cohen[17]	Bulk	Swelled to equilibrium, polymerization in the presence of excess monomer.

4.2. DEVELOPMENT OF THEORY

In order to properly analyze and compare the data in the four papers, two equations especially derived for compatible IPNs were employed. These equations relate the swelling and modulus behavior to the double-network composition and crosslink level. In all cases, a sequential mode of synthesis is assumed where network II swells network I.

4.2.1. Sequential IPN Rubbery Modulus

Let us consider the Young's modulus, E, of a sequential IPN having both polymers above their respective glass transition temperatures. A simple numerical average of the two network properties results in[23]

$$E = 3(\nu_1 v_1 + \nu_2 v_2)RT \qquad (4.1)$$

where ν_1 and ν_2 represent the number of moles of network I and II chains per cm^3, respectively, and v_1 and v_2 are the volume fractions of the two polymers, respectively. The quantities R and T stand for the gas constant and the absolute temperature, respectively. While equation (4.1) should apply to SINs,[24,25] interestingly enough, it has never been tested. Nevertheless, it does not adequately express the Young's modulus of a sequential IPN.

An equation for a sequential IPN begins with a consideration of the front factor, r_i^2/r_f^2,[26,27] where r_i represents the actual end-to-end distance of a chain segment between crosslink sites in the network, and r_f represents the equivalent free chain end-to-end distance.[21] For a single network,

$$E_1 = 3\left(\frac{\overline{r_i^2}}{\overline{r_f^2}}\right)\nu_1 RT \qquad (4.2)$$

If the network is unperturbed, the front factor is assumed to equal unity. For the case of perturbation via network swelling,

$$\left(\frac{\overline{r_i^2}}{\overline{r_f^2}}\right)_1 = \frac{1}{v_1^{2/3}} \qquad (4.3)$$

For network I swollen with network II, the concentration of ν_1 chains is reduced to $\nu_1 v_1$. Substituting this and equation (4.3) into equation (4.2), the contribution to the modulus by network I, E_1, becomes

$$E_1 = 3v_1^{1/3}\nu_1 RT \qquad (4.4)$$

With dilution of network II by network I, the contribution to the modulus by network II may be written

$$E_2 = 3v_2\nu_2 RT \qquad (4.5)$$

Equation (4.5) assumes that network II is dispersed in network I, yet retains sufficient continuity in space to contribute to the modulus. Since the chain conformation of network II undergoes a minimal perturbation, the quantity $\overline{r_i^2}/\overline{r_f^2}$ is further assumed to be unity.

The modulus contributions may be added:

$$E_1 + E_2 = E \qquad (4.6)$$

This assumes mutual network dilution and co-continuity, with no added internetwork physical crosslinks, and since the final material has only the two networks,

$$v_1 + v_2 = 1 \qquad (4.7)$$

Then E may be expressed

$$E = 3(v_1^{1/3}\nu_1 + v_2\nu_2)RT \qquad (4.8)$$

as the final result. Equation (4.8) will always yield a larger value of E than equation (4.1), because v_1 is a fractional quantity. Further, equation (4.8) does not yield a simple numerical average of the two crosslink densities.

In a somewhat related network problem, Meissner and Klier[28] derived an equation to express the behavior of supercoiled networks. Such materials may be prepared by polymerization and/or crosslinking in solution, and evaporation of the solvent.

and

$$\alpha_2^\circ = \left(\frac{1}{v_2^\circ}\right)^{1/3} = \left(\frac{V^\circ}{V^\circ - V_0}\right)^{1/3} \tag{4.18}$$

where V° is the initial (unswollen) volume of the IPN. Note that it is the ratio of α_2/α_2° which has the major physical significance, rather than the individual values, since when $\alpha_2/\alpha_2^\circ = 1$, $\Delta S_{el(2)} = 0$. Substituting $\Delta G_{el(2)} = T\Delta S_{el(2)}$, from equations (4.16) and (4.17) and expressing v_2 in moles, Thiele and Cohen[17] find

$$N_A\left(\frac{\partial \Delta G_{el(2)}}{\partial \alpha_2}\right)_{T,P}\left(\frac{\partial \alpha_2}{\partial n_s}\right)_{T,P} = -N_A\left(\frac{\partial T\Delta S_{el(2)}}{\partial \alpha_2}\right)_{T,P}\left(\frac{\partial \alpha_2}{\partial n_s}\right)_{T,P} \tag{4.19}$$

$$= \frac{RT\bar{v}_s v_2}{V^\circ - V_0}\left(v_2^{\circ 2/3} v_2^{1/3} - \frac{v_2}{2}\right) \tag{4.20}$$

Substitution of the results of equations (4.12), (4.15), and (4.20) into equation (4.11) yields

$$\mu_s - \mu_s^\circ = RT\left\{\ln[1 - (v_1 + v_2)] + (v_1 + v_2) + \chi_s(v_1 + v_2)^2 + \frac{\bar{v}_s v_1}{V_0}\left(v_1^{1/3} - \frac{v_1}{2}\right)\right.$$

$$\left. + \frac{\bar{v}_s v_2}{V^\circ - V_0}\left(v_2^{\circ 2/3} v_2^{1/3} - \frac{v_2}{2}\right)\right\} \tag{4.21}$$

In the equilibrium case, of course, $\mu_s - \mu_s^\circ = 0$. Equation (4.21) may be cast into a form more directly applicable to experiments:

$$\ln(1 - v_1 - v_2) + v_1 + v_2 + \chi_s(v_1 + v_2)^2$$

$$= -V_s v_1'(v_1^{1/3} - v_1/2) - V_s v_2'(v_2^{\circ 2/3} v_2^{1/3} - v_2/2) \tag{4.22}$$

The quantities v_1' and v_2' represent the crosslinks of the corresponding single networks in mol/cm^3 as determined separately on the single networks by equation (4.9). The following equation differs from the original Thiele–Cohen derivation[17] by the insertion[21] of the term $(1/v_0^{})^{2/3}$ in the first term on the right-hand side of equation (4.22):

$$\ln(1 - v_1 - v_2) + v_1 + v_2 + \chi_s(v_1 + v_2)^2$$

$$= -V_s v_1'(1/v_1^\circ)^{2/3}(v_1^{1/3} - v_1/2) - V_s v_2'(v_2^{\circ 2/3} v_2^{1/3} - v_2/2) \tag{4.22'}$$

This term represents the thermoelastic front factor to account for internal energy changes on swelling.[32] Typical values of the term $(1/v_1^\circ)^{2/3}$ range from 1.1 to 4.6 as v_1° varies from 0.9 to 0.1, respectively. Equation (4.22') will be denoted as the modified Thiele–Cohen equation with the insertion of the new term.

It must be emphasized that network I always dilutes network II, while network II always swells and dilutes network I. Of course, a solvent swells both networks, but to different extents.

The experimental quantity of interest is the total volume of polymer v in the swollen IPN:

$$v = v_1 + v_2 \qquad (4.23)$$

A word must be said about the use of equations (4.8) and (4.22). In each case, the modulus of the single networks determined v_1 and v_2 for equation (4.8), and the Flory–Rehner equation determined v_1' and v_2' for equation (4.22) from the equivalent single network. In this way, several effects existing in the single networks were minimized, such as physical crosslinks, incomplete crosslinking, and branching. Thus, differences from theory emphasize new effects due to sequential IPN formation.

Further analytical expressions for the properties of more or less homogeneous IPNs and SINs will be found in Section 6.7.2.1.

4.3. RELATIVE NETWORK CONTINUITY AND PHYSICAL CROSSLINKS

4.3.1. Swelling Studies

Equations (4.22) and (4.22′) are predicated on the assumptions that both networks are continuous in space, network II swells network I, and that the swelling agent then swells both networks. The several PS/PS IPNs under consideration constitute an excellent model system with which to examine fundamental polymer parameters. Questions of interest in the field of IPNs relate to the relative continuity of networks I and II and their consequent relative contribution to physical properties and the extent of formation of physical crosslinks or actual chemical bonds between the two networks. If equations (4.22) and (4.22′) are obeyed exactly, the implicit assumptions require that both networks be mutually dissolved in one another and yet remain chemically independent. Then the only features of importance are the crosslink densities and the proportions of each network.

Values for v's obtained via equations (4.22) and (4.22′) vs. v's determined from the swelling of the IPNs in toluene[14-17] are shown in Figure 4.1. While both the modified and unmodified equations (4.22) and (4.22′) fit the data, the modified equation fits somewhat better.[21,22]

If new physical or chemical crosslinks were added during IPN formation, one would expect the data to be shifted to the right of the theoretical line. Figure 4.1 indicates that substantially no new physical or chemical

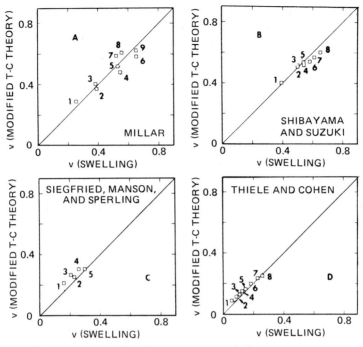

Figure 4.1. Equilibrium swelling of polystyrene/polystyrene homo-IPNs by toluene. Swelling values predicted by the modified Thiele–Cohen equation (4.22′) vs. experimental swelling values.[21] The following pairs of numbers indicate, for each sample: Vol. %DVB in network I/vol. %DVB in network II, and % network I/% network II. A: (1) 1/1, 21/79; (2) 2/1, 32/68; (3) 2/2, 32/68; (4) 4/1, 44/56; (5) 4/4, 44/56; (6) 7/1, 56/44; (7) 2/10, 32/68; (8) 4/10, 44/56; (9) 7/7, 56/44. B: (1) 1.9/1.9, 31/69; (2) 3.1/5.3, 37/63; (3) 2.2/7.2, 33/67; (4) 4.3/4.3, 44/56; (5) 5.3/2.2, 49/51; (6) 5.2/3.1, 49/51; (7) 7.2/2.2, 53/47; (8) 7.2/7.2, 53/47. C: (1) 0.22/2.2, 75/25; (2) 0.22/2.2, 55/45; (3) 0.22/2.2, 50/50; (4) 0.22/2.2, 25/75; (5) 0.22/2.2, 25/75. D: (1) 0.3/0.3, 5/95; (2) 0.3/0.6, 5/95; (3) 0.3/1.0, 5/95; (4) 0.7/0.3, 10/90; (5) 0.7/0.6, 10/90; (6) 0.7/1.5, 10/90; (7) 1.5/0.6, 19/81; (8) 1.5/1.5, 19/81.

crosslinks are present, at least when the networks are fully swollen. However, analysis of the data does indicate that the limitation imposed on swelling by network I is out of proportion to its crosslink density and proportion in the IPN. The clearest indication of the conclusion can be found by examining samples 4, 5, and 8 in Figure 4.1A[15]. The major variable is the crosslink density of network II. Note that the experimental v's (x axis) substantially do not vary in this series, while both the Thiele–Cohen[17] and modified Thiele–Cohen equations predict substantial variations. One way of viewing the data is that it does not matter as much what the crosslink level of network II is relative to network I. The other series in Figure 4.1A substantiate this finding, albeit less dramatically. A statistical analysis of

Table 4.2. IPN Swelling Factors Unaccounted for by Theory[21]

Investigator	Added physical crosslinks?	Network I domination?
Millar[15]	No	Yes
Shibayama and Suzuki[16]	No	Slight
Siegfried, Manson and Sperling[14]	No	No
Thiele and Cohen[17]	No	No

Figure 4.1B data indicates a similar trend. Neither the data of Thiele and Cohen Figure 4.1D[17] nor the data of Siegfried, Manson, and Sperling[14] Figure 4.1C support this conclusion. The results of this analysis are summarized in Table 4.2.

4.3.2. Analysis of Modulus Data

Figures 4.2, 4.3, and 4.4 show the modulus predicted by equation (4.8) vs. Young's modulus, E (experiment). As with the swelling data, the network imperfections and the contributions of the physical crosslinks, if any, were minimized by determining the two crosslink levels required for E (theory) on the separate homopolymer networks. Unfortunately, Millar did not report modulus data for his polystyrene/polystyrene homo-IPNs.

Data lying to the right of the theoretical line provide an indication of physical crosslinks, since the experimental modulus is larger than the theoretical modulus. A vertical stacking of points containing the same network I crosslink density (but different network II's), suggests the dominance of network I over network II, since the experimental modulus is relatively constant.

The data, unfortunately, are observed to be erratic and inconsistent with one another. In Shibayama and Suzuki's data, Figure 4.2, the preponderance of data to the right of the theoretical diagonals suggest added physical crosslinks. Domination of network I over network II is suggested by a slight vertical stacking noted in the samples numbered 2 and 3. More evident is the horizontal stacking caused by inverting networks I and II; note the 3,7 pair and 6,9 pair.

Siegfried, Manson, and Sperling's data (Figure 4.3) show the data to the left of the line, vertically stacked. (All of the polymer network I compositions have the same crosslink density.) Hence network I is assumed to dominate network II.

In Figure 4.4, Thiele and Cohen's data fit the theory surprisingly well, however, with a slight indication of vertical stacking.

Figure 4.2. Young's modulus behavior of PS/PS homo-IPNs. Shibayama and Suzuki's data[16] plotted according to equation (4.8) for E (theory) vs. experiment.[23] (1) 1.9/1.9, 31/69; (2) 3.1/3.1, 37/63; (3) 3.1/5.3, 37/63; (4) 4.3/4.3, 44/56; (5) 5.3/2.2, 49/51; (6) 2.2/7.2, 33/67; (7) 5.3/3.1, 49/51; (8) 5.3/5.3, 49/51; (9) 7.2/2.2, 53/47; (10) 7.2/7.2, 53/47.

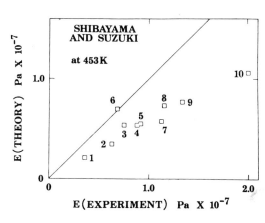

Figure 4.3. Siegfried et al.'s data[14] plotted according to equation (4.8). Note vertical stacking of data.[23] (1) 0.22/2.2, 75/25; (2) 0.22/2.2, 55/45; (3) 0.22/2.2, 50/50; (4) and (5) 0.22/2.2, 25/75.

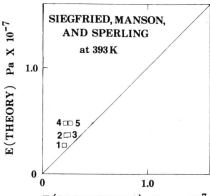

Figure 4.4. Data of Thiele and Cohen,[17] plotted according to equation (4.8).[22] (1) 0.2/0.2, 3/97; (2) 1.0/0.2, 13/87; (3) 0.6/0.6, 9/91; (4) 0.2/1.0(NIS), 3/97; (5) 0.2/1.0, 3/97; (6) 1.0/1.0(NIS), 13/87; (7) 1.0/1.0, 13/87.

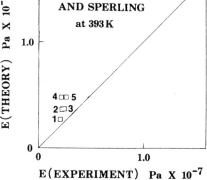

The effects of network I domination and added physical crosslinks are summarized in Table 4.3. Unfortunately, Millar[15] presented no modulus data as his materials were suspension-sized particles.

Before any discussion of the results, two comments must be made. First, a much greater scatter of the data was encountered for the modulus data, compared to the swelling data. Second, in some cases, the values of the modulus were below those predicted, for single networks and IPNs alike. Because the data were collected over a range of temperatures (all above 100°C), it was more convenient to work in terms of the crosslink levels rather than in terms of the moduli. From equation (4.8), a value of

$$\nu \text{ (theory)} = v_1^{1/3} \nu_1 + v_2 \nu_2 \qquad (4.24)$$

was defined. [In itself, ν (theory) is an effective crosslink density and not representative of a measurable number of crosslinks.]

If the experimental values of ν from $E = 3\nu RT$ exceeded ν (theory), this was taken as evidence for new physical crosslinks caused by IPN formation. Likewise, a slower than expected variation within a composition series or near series (via statistical analysis) was taken as evidence for an outsized contribution by network I.

While Table 4.3 affirms network I domination, there appears to be conflicting evidence for added physical crosslinks.

In addition to the swelling and modulus data analyzed above, the creep data and morphology (electron microscopy) studies by Siegfried et al.[14] point to a greater continuity of network I. In the language of Lipatov and Sergeeva,[12] network II appears to behave somewhat like a filler for network I. This is all the more surprising in the present case, where both polymers are identical in composition.

It should be pointed out that both viscoelastic data and transmission electron microscopy data by Siegfried et al.,[14] not analyzed above, also

Table 4.3. IPN Modulus Factors Unaccounted for by Theory[22]

Investigator	Added physical crosslinks?	Network I domination?
Millar[15]	N.A.[a]	N.A.[a]
Shibayama and Suzuki[16]	Yes	Yes
Siegfried, Manson, and Sperling[14]	No	Yes
Thiele and Cohen[17]	No	Very slight

[a] N.A. = Not applicable.

support the speculation that network I dominates network II. In fact, it is seen from the electron micrograph in Figure 4.5,[14] that network I apparently exhibits greater continuity in space. Network II is seen to form small domains about 75 Å in diameter within network I.

Overall, network I domination is apparently a real phenomenon, but is much more obvious in the solid state than in the swollen state. Perhaps the networks behave much more "ideally" when swollen, which is not surprising.

The idea of network I dominating the properties of the IPN through its greater continuity in space has some important implications in thermoset resin synthesis, such as epoxy materials. The suggestion is that if a mixture of monomers is simultaneously polymerizing, then that portion of the material already incorporated in the network at the time of gelation may tend to be more continuous in space and dominate the properties. That material polymerized later in time, statistically, may form less continuous domains

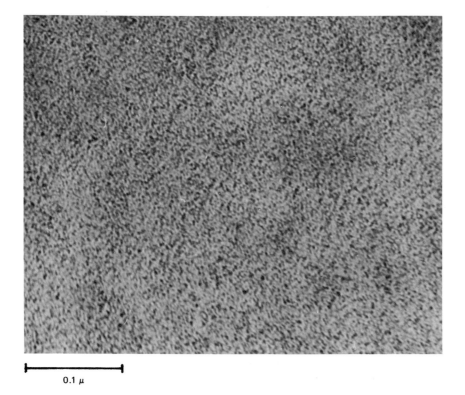

0.1 μ

Figure 4.5. Morphology of 50/50–0.4/4% DVB + 1% isoprene homo-IPN. Polymer network II, darker regions stained with OsO_4, appear as domains near 75 Å in diameter.[14]

and act like a filler to a greater or lesser extent. Partial confirmation of this concept has already been obtained.[33,34]

While there is some evidence for new physical crosslinks, unfortunately the data do not permit a reasonable conclusion either way. Only the modulus data of Shibayama and Suzuki, Figure 4.2, indicate the presence of any added physical crosslinks contributing to the rubbery modulus. It must be emphasized that Siegfried *et al.*'s[14] data, Figure 4.3, indicate fewer physical crosslinks in the homo-IPN than in the corresponding single networks, because the data lie to the left of the line.

Flory has concluded[35] that physical crosslinks should not contribute to the rubbery modulus. Therefore, added physical crosslinks arising through interpenetration should also not contribute. So then the modulus should not reflect the number of the physical crosslinks. However, other workers[2-5] find physical crosslinks making important contributions, if through various routes. Clearly more work is required before the problem will be resolved.

REFERENCES

1. P. J. Flory, *Proc. R. Soc. London Ser. A.* **351**, 351–380 (1976).
2. O. Kramer, V. Ty, and J. D. Ferry, *Proc. Natl. Acad. Sci. USA* **69**, 2216 (1972).
3. R. L. Carpenter, O. Kramer, and J. D. Ferry, *Macromolecules* **10**, 117 (1977).
4. P. J. Flory, *Trans. Faraday Soc.* **56**, 772 (1960).
5. O. Kramer, in Europhys. Conference Abstracts, 3C, *Structure and Properties of Polymer Networks*, European Physical Society, Warsaw, Poland, April (1979).
6. W. W. Graessley, *Adv. Polym. Sci.* **16** (1974).
7. H. M. James and E. Guth, *J. Chem. Phys.* **15**, 669 (1947).
8. B. N. Kolarz, *J. Polym. Sci.* **47**, 197 (1974).
9. W. Trochimczuk, in Europhysics Conference Abstracts, 3C, *Structure and Properties of Polymer Networks*, European Physical Society, Warsaw, Poland, April (1979).
10. F. G. Hutchinson, R. G. C. Henbest, M. K. Leggett, U.S. Pat. 4,062,826 (1977).
11. H. L. Frisch, K. C. Frisch, and D. Klempner, *Mod. Plast.* **54**, 76, 84 (1977).
12. Yu. S. Lipatov and L. M. Sergeeva, *Russian Chem. Rev.* **45**,(1), 63 (1976).
13. D. A. Thomas and L. H. Sperling, in *Polymer Blends*, D. Paul and S. Newman, eds., Academic, New York (1978), Chap. 11.
14. D. L. Siegfried, J. A. Manson, and L. H. Sperling, *J. Polym. Sci., Polym. Phys. Ed.* **16**, 583 (1978).
15. J. R. Millar, *J. Chem. Soc.* 1311 (1960).
16. (a) K. Shibayama and Y. Suzuki, *Kobunshi Kagaku* **23**, 24 (1966); Reprinted in (b) *Rubber Chem. Technol.* **40**, 476 (1967).
17. J. L. Thiele and R. E. Cohen, *Polym. Eng. Sci.* **19**, 284 (1979).
18. H. A. Clark, U.S. Pat. 3,527,842 (1970).
19. J. J. P. Staudinger and H. M. Hutchinson, U.S. Pat. 2,539,377 (1951).
20. Yu. S. Lipatov, V. F. Rosovizky, and V. F. Babich, *Eur. Polym. J.* **13**, 651 (1977).
21. D. L. Siegfried, D. A. Thomas, and L. H. Sperling, *Macromolecules* **12**, 586 (1979).

22. D. L. Siegfried, D. A. Thomas, and L. H. Sperling, in *Polymer Alloys II*, K. C. Frisch and D. Klempner, eds., Plenum (1980).
23. K. J. Smith and R. J. Gaylord, *J. Polym. Sci., A-2*, **10**, 283 (1972).
24. S. C. Kim, D. Klempner, K. C. Frisch, W. Radigan, and H. L. Frisch, *Macromolecules* **9**, 258 (1976).
25. N. Devia-Manjarres, J. A. Manson, L. H. Sperling, and A. Conde, *Polym. Eng. Sci.* **18**, 200 (1978).
26. A. V. Tobolsky and M. C. Shen, *J. Appl. Phys.* **37**, 1952 (1966).
27. M. C. Shen, T. Y. Chen, E. H. Cirlin, and H. M. Gebhard, in *Polymer Networks: Structure and Mechanical Properties*, A. J. Chompff and S. Newman, eds., Plenum, New York (1971).
28. B. Meissner and I. Klier, in Europhysics Conference Abstracts, 3C, *Structure and Properties of Polymer Networks*, European Physical Society, Warsaw, Poland, April (1979).
29. P. J. Flory and J. Rehner, *J. Chem. Phys.* **11**, 512 (1943).
30. P. J. Flory, *Principles of Polymer Chemistry*, Cornell University, Ithaca, New York (1953).
31. J. J. Hermans, *Trans. Faraday Soc.* **43**, 591 (1947).
32. A. V. Galanti and L. H. Sperling, *Polym. Eng. Sci.* **10**, 177 (1970).
33. S. C. Misra, J. A. Manson, and L. H. Sperling, *Am. Chem. Soc. Div. Org. Coat. Plast. Chem. Prepr.* **39**(2), 146 (1978).
34. S. C. Misra, J. A. Manson, and L. H. Sperling, *Am. Chem. Soc. Div. Org. Coat. Plast. Chem. Prepr.* **39**(2), 152 (1978).
35. P. J. Flory, in Europhysics Conference Abstracts, 3C, *Structure and Properties of Polymer Networks*, European Physical Society, Warsaw, Poland, April (1979).

SYNTHESIS OF IPNs AND RELATED MATERIALS

5.1. INTRODUCTION

Depending on the needs of the investigator, the synthesis of an IPN may be considered from several different points of view. Description according to chemical composition, synthetic mode, and final topology each have their place. Any method of describing the synthesis must distinguish the IPNs from other multicomponent systems and show the relationships among the IPNs. The scope and limitations of each descriptive mode first require definition.

1. *Chemical Composition.* A list of the several reactants such as monomers, prepolymers, crosslinkers, and initiators will be provided, together with a step by step reaction scheme. Such a presentation provides the greatest detail, yet by itself only provides an empirical formulation, lacking greatly in capability of systematization.

2. *Synthetic Mode.* Two lines of thought are encompassed by the term "synthetic mode": (a) the time order of the reactions, principally sequential or simultaneous, and (b) the mixing mode, related to the use of latexes, mutual solutions, stirring, swelling techniques, etc. Both (a) and (b) are helpful in determining the relationships among the different compositions, and provide hints concerning the morphology and properties.

3. *Topology.* This mode of classification is concerned with the spatial arrangements of the chains and answers questions relating to the presence or absence of crosslinking, grafting, etc. An understanding of the product structure sometimes requires a knowledge of the time sequence of the reactions [see 2(a) above]. The topology is described by the nomenclature scheme illustrated in Chapter 3.

Elements of both points 2 and 3 above require an understanding of each chemical reaction and side reaction, and all of the physical transformations taking place—an understanding that is seldom achieved. In cases of doubt, the intent of the original investigator will be mentioned, along with other possible interpretations. In the following, elements of each method will be presented, as necessary, to convey to the reader an accurate picture.

5.2. SEQUENTIAL IPNs*

The term "sequential" refers to the time order of polymerization.[1–5] Polymer network I may be formed by simultaneous polymerization and crosslinking, through the use of multifunctional crosslinkers such as divinyl benzene, or polymer I can be crosslinked after its initial preparation, as in the vulcanization of rubber. Monomer II, together with requisite cross-linkers and initiators, is swollen into polymer network I, and polymerized *in situ*. Table 5.1 summarizes the types of homopolymer networks employed.

5.2.1. Poly(ethyl acrylate)/polystyrene, PEA/PS

In the following discussion poly(methyl methacrylate), PMMA, or P(S-*co*-MMA) replaced PS as polymer II, where desired.[5]

All of the IPNs were synthesized by photopolymerization techniques although thermally induced reactions with peroxide work as well. A monomer containing dissolved benzoin as initiator and tetraethylene glycol dimethacrylate (TEGDM) as the crosslinking agent was polymerized by exposure to ultraviolet (uv) light. The composition throughout was 0.5 ml of TEGDM and 0.3 g of benzoin per 100 ml of monomer, which was ethyl acrylate, methyl methacrylate, or styrene. After vacuum drying network I to constant weight to remove the unreacted monomer (usually less than 2% weight loss), it was swollen with monomer II solution. The duration of imbibing depended on the desired overall composition. The monomer II was then polymerized *in situ* after uniformity of composition had been attained through diffusion. This was followed by a second drying step. Several samples were also exposed to swelling and extraction studies. As expected, the materials swelled but did not dissolve, confirming the network charac-teristics. Only 1–2% extractables by weight were observed.

Four series of IPNs were polymerized, the compositions of which are given in Table 5.2. The underlined polymer was polymerized first. This was always the elastomer PEA (normal IPNs), except for series I (inverse series), where the plastic homopolymer PS or PMMA was polymerized first. The B in PEAB indicates that the PEA contained 1% butadiene as a comonomer to permit staining for electron microscopy. The letters E, L, P, and I denote elastomeric, leathery, plastic, and inverse series, respectively. In composi-tions containing both S- and MMA-mers, a random copolymer was formed with the indicated composition. The actual compositions employed can be portrayed with the aid of a pseudoternary phase diagram, as shown in Figure 5.1 for the normal IPNs. Only the border compositions (no random

---

* Topological designation: $C_1 \, O_I \, C_2$.

Table 5.1. Typical Sequential IPN Components

Monomer	Crosslinker	Initiator	Polymerization mode	Remarks	Reference
Polymer I (usually an elastomer)					
Ethyl acrylate	Tetraethylene glycol dimethacrylate (TEGDM)	Benzoin	uv photopolymerization in bulk	Thermal polymerization with dicumyl peroxide feasible.	1
Butadiene	Cast film heated with peroxides	Potassium persulfate	Emulsion	Difficult to synthesize cross-linked polybutadiene in bulk.	2
Castor oil	Tolylene diisocyanate (TDI)	None	Bulk	Castor oil contains three hydroxyl groups per molecule.	3
Polymer II (usually a plastic)					
Styrene	Divinyl benzene	Benzoin	uv photopolymerization in bulk	Usually requires 1–3 days. Thermal polymerization feasible.	1–3
Methyl methacrylate	TEGDM	Benzoin	uv photopolymerization	Isomeric with ethyl acrylate.	1

1. V. Huelck, D. A. Thomas, and L. H. Sperling, *Macromolecules* **5**, 340, 348 (1972).
2. A. A. Donatelli, L. H. Sperling, and D. A. Thomas, *Macromolecules* **9**, 671, 676 (1976).
3. G. M. Yenwo, J. A. Manson, J. Pulido, L. H. Sperling, A. Conde, and N. Devia, *J. Appl. Polym. Sci.* **21**, 153 (1977).

Table 5.2. Composition of IPNs[5]

Series	Polymer	Wt % polymer			
		1	2	3	4
E	PEAB	74.4	75.9	75.5	72.2
	P(S-co-MMA)	25.6–0	15.7–8.4	9.9–14.6	0–27.8
L	PEAB	48.8	51.2	48.4	47.1
	P(S-co-MMA)	51.2–0	23.4–25.4	13.6–38.0	0–52.9
P	PEAB	23.9	24.7	25.4	23.3
	P(S-co-MMA)	76.1–0	40.9–34.4	24.9–49.7	0–76.7
I	PS	24.6	50.7	71.4	0
	PMMA	0	0	0	77.5
	PEAB	75.4	49.3	28.6	22.5

copolymers) were investigated in the inverse materials. The "normal" and "inverse" notation is arbitrary, and depended only on inverting the time order of polymerization.

Series L, which contains approximately 50 wt % PEAB, provides leathery materials, while series E, which contains 75 wt % PEAB, yields self-reinforcing elastomeric materials. Series P contains only approximately 25 wt % PEAB and exhibits properties of toughened plastics. Since polymer I in an IPN is always strained owing to the diffusional swelling forces of monomer II (see Chapter 4), it was of particular interest to prepare an

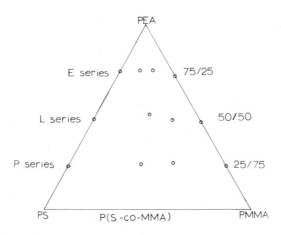

Figure 5.1. Ternary composition diagram showing materials. Note that styrene and methyl methacrylate compositions are random copolymers of the two.[5]

inverse series of IPNs (series I) where the plastic component was polymerized first. Since the compositions of this series are almost identical to
some of the border compositions (containing only one plastic-forming mer)
of the L, E, and P series, they may be effectively compared with those. The
following samples match in overall composition (see Table 5.2): I1/E1,
I2/L1, I3/P1, I4/P4.

The reasons underlying the selection of the three monomers involved
should be mentioned. PEA and PMMA are chemically isomeric, and hence
expected to be much more compatible with each other than the PEA/PS
pairs, which are known to be incompatible (see Chapter 6). On the other
hand, PMMA and PS have glass–rubber transitions of 105 and 100°C,
respectively. As a result, their random copolymers will have essentially the
same glass-transition temperatures (iso-T_g) as the plastic homopolymers.
This last reason will simplify the interpretation of the mechanical results to
be presented in Chapter 6.

The several compositions prepared in the above manner can be related
easily through the ternary diagram shown in Figure 5.1.

5.2.2. SBR/PS

Styrene–butadiene rubber, SBR, served as polymer I. The IPNs and
semi-IPNs were synthesized by thermal polymerization techniques.[6,7] The
SBR for the semi-IPNs of the first kind and for the full IPNs was prepared by
dissolving the rubber in benzene, adding the appropriate amount of dicumyl
peroxide (Dicup) for crosslinking, and then evaporating the solvent. The
SBR was then cured in a compression molding operation at a temperature of
325°F and at a pressure of 40–50 psi for 45 min. See Table 5.3 for levels of
crosslinker employed.

In order to form the plastic phase of the semi-IPNs of the first kind and
the full IPNs, styrene monomer solutions were prepared containing 0.4%
(w/v) dicumyl peroxide and the several quantities of divinylbenzene shown
in Table 5.3. To swell in the monomer, a known weight of crosslinked rubber
was immersed in the solution at ambient conditions. The ratio of styrene to
divinylbenzene actually imbibed was not determined. The duration of
imbibing was dependent upon the desired final IPN composition. The
swollen polymer then was placed in an airtight container with a saturated
styrene atmosphere for approximately 12 hr so that a uniform distribution of
monomer could be achieved throughout the sample. Next, the styrene was
polymerized thermally at 50°C for a period of 4 days and at 100°C for 1 hr.
Finally, the IPN was subjected to a vacuum-drying operation to remove any
unreacted monomer. Semi-IPNs of the second kind were prepared by

Table 5.3. Composition of SBR/PS Semi-IPNs and Full
IPNs[7]

Series no.	% Dicup in SBR, w/w	%DVB in PS, v/v	Wt. % S in SBR
1	0	1.3	5
2	0	2.0	5
3	0.10	0	5
4	0.10	1.0	5
5	0.10	2.0	5
6	0.20	0	5
7	0.20	1.3	5
8	0.20	2.0	5
9	0	2.0	23.5
10	0.05	0	23.5
11	0.05	2.0	23.5
12	0.10	0	23.5
13	0.10	2.0	23.5
14	0.20	0	23.5
15	0.20	2.0	23.5
16	0.00	0.0	5

dissolving the uncrosslinked SBR into a styrene monomer solution followed
by a thermal polymerization similar to the other IPNs. Sixteen series of IPNs
were prepared with the variables manipulated as shown in Table 5.3.

5.2.3. Castor Oil/Polystyrene

As shown in structure (5.1), castor oil has three hydroxyl groups per
molecule, which were used to form the polyurethane[8–10]:

$$
\begin{array}{l}
\text{H}_2\text{C}-\text{O}-\overset{\text{O}}{\overset{\|}{\text{C}}}-(\text{CH}_2)_7-\text{CH}{=}\text{CH}-\text{CH}_2-\overset{\text{OH}}{\overset{|}{\text{CH}}}-(\text{CH}_2)_5-\text{CH}_3 \\[2mm]
\text{HC}-\text{O}-\overset{\text{O}}{\overset{\|}{\text{C}}}-(\text{CH}_2)_7-\text{CH}{=}\text{CH}-\text{CH}_2-\overset{\text{OH}}{\overset{|}{\text{CH}}}-(\text{CH}_2)_5-\text{CH}_3 \quad (5.1) \\[2mm]
\text{H}_2\text{C}-\text{O}-\overset{\text{O}}{\overset{\|}{\text{C}}}-(\text{CH}_2)_7-\text{CH}{=}\text{CH}-\text{CH}_2-\overset{\text{OH}}{\overset{|}{\text{CH}}}-(\text{CH}_2)_5-\text{CH}_3
\end{array}
$$

Castor oil–urethane elastomers were prepared by reacting 2, 4-tolylene
diisocyanante, TDI, 80/20:2, 4/2, 6 TDI, or hexamethylene diisocyanate,
HDI, with castor oil. The last reaction was rather slow and thus dibutyltin

dilaurate, 0.001 g per gram of HDI, was used as a catalyst. Since TDI hydrolyzes significantly in the presence of trace amounts of water, DB-grade castor oil was employed.

The reaction between TDI and castor oil is exothermic and bubbles are produced in the reaction mixture. Castor oil contains a few tenths of a percent volatile material that will evaporate as the temperature of the reaction mixture is increased. Some of the bubbles produced are trapped in the mixture as the viscosity increases. (Stirring with a Teflon-coated magnetic spin bar also produces some bubbles.) In order to produce elastomers that are bubble free, the reaction is carried out in two stages.

Stage I. A known weight of castor oil was mixed with excess TDI (the excess here refers to the ratio of NCO groups to OH groups being larger than 1.0) at room temperature to produce an isocyanate-terminated prepolymer. The mixture was stirred vigorously for at least 1 hr. The bubbles present were removed by applying a vacuum to the prepolymer for about 15 min. This resulted in a clear, bubble-free, highly viscous liquid.

Stage II. In this stage the prepolymer is crosslinked with excess castor oil. The degassed prepolymer was mixed with enough DB castor oil to give a final predetermined NCO/OH ratio.

In order to synthesize the IPN, the urethane elastomer was swelled with styrene containing 0.4% benzoin as initiator and 1% divinylbenzene (DVB) as crosslinker. Polymerization of the styrene was carried out by ultraviolet radiation at room temperature for 24 hr. The corresponding castor oil–polyester/polystyrene SINs are discussed in Section 5.5.

Several remarks of a general nature are in order.

1. The syntheses illustrated above make tough plastics, reinforced elastomers, or leathery materials suitable for broad-temperature-damping compositions, depending on the ratio of plastic to rubber, and extent of molecular mixing. Properties will be discussed in Chapter 6 and 7.

2. The crosslink level in polymer I limits the extent of swelling possible with monomer II. The Flory–Rehner equation, (4.9), provides a quantitative basis for estimating this limit, if the crosslink level is known. However, less than equilibrium amounts of monomer II are frequently employed.

3. Polymer network I is usually an elastomer because of ease and rapidity of swelling. If network I is below its T_g, extreme care must be exercised to prevent solvent-type stress cracking.

4. Sheets of 3–5 mm thickness may be prepared using a glass plate sandwich technique. Heat transfer becomes a problem with very thick specimens.

5. Both chain and stepwise polymerization modes have been used to form polymer I. However, only limited studies have been done using a stepwise polymerization for polymer II.

6. The chief disadvantage of a sequential IPN resides in its thermoset nature, providing processing difficulties.

5.2.4. Filled IPNs

Lipatov *et al.*[11,12] prepared a series of sequential IPNs, sometimes adding Aerosil (silica) filler to the material. The first polymer was a polyurethane, prepared from polyethyleneglycol adipate (mol wt 1800–2000), and an adduct of trimethylolpropane with tolylene diisocyanate. The requisite quantity of Aerosil was mixed in, and the mass allowed to react. Styrene and divinyl benzene were then swollen in, and thermally polymerized with benzoyl peroxide.

5.2.5. Anionically Polymerized IPNs

Nearly all of the materials described in this monograph were prepared by some combination of free-radical and/or condensation reactions. Most recently, however, Lipatova *et al.*[13] obtained grafted sequential IPNs on the basis of matrices from "living" network polymers, using anionic polymerizations for both networks I and II.

Monomers α,ω-dimethacrylbis (triethylene glycol) phthalate (MGP) or trioxyethylene α,ω-dimethacrylate (TMA) were polymerized with Na-naphthalene. Styrene and divinyl benzene were swollen in, and the reaction continued. Network I is densely crosslinked under these conditions.

5.3. LATEX INTERPENETRATING ELASTOMERIC NETWORKS (IENs)

This synthesis, undertaken by Klempner, Frisch, and Frisch[14–18] makes use of two noninteracting types of latexes. These are synthesized separately as linear polymers, mixed together, along with crosslinkers and catalysts, and co-coagulated. The material is then subjected to a simultaneous crosslinking step via heating.* Table 5.4 delineates materials employed in the synthesis.[14] One latex usually was a polyurethane.

The materials are designated as IENs because both latexes are normally elastomeric and to help distinguish them from other types of IPNs.

---

* Topological designation:

$$\begin{pmatrix} P_1 \\ O_M \\ P_2 \end{pmatrix} O_C \begin{pmatrix} P_1 \\ P_2 \end{pmatrix}$$

Table 5.4. Materials for IEN Synthesis[18]

Material	Designation	Description	Supplier
Poly(urethane-urea)	U-1033 U-E-503	50% water emulsions of crosslinked polymers based on tolylene diisocyanate and a blend of poly(oxypropylene) glycols and triols and chain extended with a diamine.[12]	Wyandotte Chemicals Corp.
Polyacrylate	H-120 H-138	Hycar latices 2600×120 and 2600×138. 50% emulsions of polyacrylates.	B. F. Goodrich Chemical Co.
Poly(2-chloro-1,3-butadiene) (polychloroprene)	N	Neoprene latex 842-A. 50% water emulsion of linear polymer of 2-chloro-1,3-butadiene.	E. I. du Pont de Nemours & Co.
Poly(styrene-butadiene)	SBR-5362	Pliolite latex 5362.69. 8% water emulsion of linear random copolymer of styrene and butadiene.	Goodyear Chemical Co.
	SBR-880	48% water emulsion of a linear copolymer of styrene and butadiene in a 52:45 ratio plus 6 parts of carboxylic acid present as acrylic. It crosslinks via the carboxyl groups by anhydride formation.	Dow Chemical Co.
Poly(dimethyl siloxane) (Silicone)	S	Silicone latex 22.40% water emulsion of hydroxyl-terminated poly(dimethyl siloxane) containing a small amount of methyl siloxane units.	Dow Corning Corp.
Stannous octoate		20% water emulsion	Dow Corning Corp.
Sulfur		68% water dispersion	R. T. Vanderbilt
Zinc oxide		60% water dispersion	R. T. Vanderbilt
Butylated bisphenol A		65% water dispersion	R. T. Vanderbilt
Zinc dibutyl dithio-carbamate			R. T. Vanderbilt
Sodium dibutyl dithiocarbamate			E. I. du Pont de Nemours & Co.

Klempner, Frisch, and Frisch describe the syntheses of both the homo-polymer networks and the IENs[18]:

1. *Individual Networks—Poly(urethane-urea).* The polymers were self-crosslinking owing to the presence of triols already in the latex. Films were cast on glass with a doctor blade, dried for 15 min at room temperature, 15 min at 80°C (or until completely tack free), and cured for 30 min at 120°C. They were removed from the glass plate by immersion in hot, distilled water. They were dried further at 80°C under a vacuum of 2 Torr.

Polyacrylates. To 200 g of the emulsion were added 1.1 g of a sulfur dispersion, 5 g of a zinc oxide dispersion, 2 g of a 50% water solution of zinc dibutyl dithiocarbamate, and 3.08 g of butylated bisphenol A dispersion. The mixture was vigorously stirred for 1 hr and films were cast on glass as above after the stirred-in air bubbles had disappeared. The films were dried and cured as above. The polymer was crosslinked by the usual vulcanization mechanism in which sulfur and zinc oxide form the crosslinks. Zinc dibutyl dithiocarbamate served as an accelerator, and the butylated bisphenol A was an antioxidant added to prevent degradation during cure.

Polychloroprene. To 200 g of emulsion were added 2 g of a 50% water solution of sodium dibutyl dithiocarbamate, 8.3 g of zinc oxide dispersion, and 2 g of a butylated bisphenol A dispersion. The mixture was stirred and films were cast, dried, and cured as above, thereby crosslinking the polymer via two mechanisms: the usual zinc oxide vulcanization, and a bisalkylation in which the crosslinking takes place at sites where there are tertiary allylic chlorine atoms formed by 1, 2 polymerization of chloroprene monomer.

Poly(styrene-butadiene). SBR-5362: To 143 g of emulsion were added 8.3 g of a zinc oxide dispersion, 3.7 g of a sulfur dispersion, 3.1 g of a butylated bisphenol A dispersion, and 2 g of a 50% by weight water solution of zinc dibutyl dithiocarbamate. The mixture was stirred, films cast, dried, and cured as above. The crosslinking mechanism again is the usual vul-canization. SBR-880: No additives were necessary since the material was self-crosslinking and is supplied with all necessary stabilizers. Films were cast, dried, and cured as above.

Silicone. To 125 g of silicone emulsion was added 25 g of stannous octoate emulsion. The mixture was stirred, films cast, dried, and cured as before, thus crosslinking the polymer through formation of siloxane bonds between chains by reaction of stannous octoate with the active hydrogen atoms present.

2. *Interpenetrating Polymer Networks, IEN Form.* Two of the above latexes were mixed together, along with crosslinking agents and catalysts. Usually the poly(urethane-urea) latex was mixed with one of the others. All components were vigorously stirred for 1 hr to yield macroscopically homogeneous mixtures.

This was followed by casting a film, drying, and raising the temperature to induce crosslinking. For example, the urethane polymer was crosslinked thermally by reaction with triols, while the polyacrylate was essentially crosslinked through sulfur and double bonds by a free-radical process. In this manner, grafting reactions between the two polymers were minimized.

5.4. LATEX IPNs

Latex IPN refers to a material where both networks appear on a single latex particle.* Synthesized by sequential polymerization, a core–shell structure is built up. A crosslinked polymer I in latex form is used as the seed latex, and upon the addition of monomer II, crosslinker, and activator (but no new soap, to discourage the formation of new particles), followed by polymerization of monomer II, the IPN is formed. Then each latex particle, ideally, consists of two crosslinked polymer molecules.[19]

5.4.1. Methacrylic/Acrylic Compositions

Methacrylate/acrylate latex IPNs were synthesized by a modified emulsion polymerization technique.[19] To 250 ml deionized, deaerated, stirred water at 60°C, 50 ml 10% (w/v) lauryl sodium sulfate solution was added, followed by 5 ml 5% (w/v) potassium persulfate solution. The calculated quantity of monomer I containing 0.4% tetraethylene glycol dimethacrylate (TEGDM) as a crosslinking agent was added at a rate of approximately 2 ml/min. When the first monomer was completely added, a minimum time of 1 hr was allowed for completion of the polymerization. Then a second 5 ml of the potassium persulfate solution was added followed by monomer II, which also contained 0.4% TEGDM. The same reaction conditions as above were followed. The combined amounts of networks I and II were 30% solids in the final latex. This method is referred to as the dropwise addition method.

When bulk addition of monomer was used, the procedure was to add all the monomer plus crosslinking agent to the flask before adding the potassium persulfate free-radical source. Although temperature control was more difficult, the maximum deviation in temperature from 60°C was usually 2°C.

Another typical IPN pair was poly(ethyl methacrylate) for polymer I, and poly(n-butyl acrylate) for polymer II. However, all acrylic monomers, styrene, and other latex-forming monomers may be employed. If polymer II

---

* Topological designation: $C_1 O_I C_2$.

is reasonably elastomeric, good films can be made. A number of materials suitable for noise- and vibration-damping applications were made by this method.[19–26]

5.4.2. PVC/Nitrile Rubber

Very recently, Sionakidis *et al.*[27] prepared a poly(vinyl chloride)/poly(butadiene-*co*-acrylonitrile) 50/(25-*co*-25) latex IPN. This particular composition was selected because the two polymers were known to be semicompatible.[28]

The latex IPNs were synthesized by a two-stage emulsion polymerization technique. The first stage consisted of making a seed latex of crosslinked PVC polymer I and then introducing the monomer II mixture of butadiene and acrylonitrile and crosslinker followed by a second polymerization. The recipe for the seed latex was as follows:

Deionized H_2O	115 ml
Sodium lauryl sulfate (SLS) emulsifier	0.3 g
Vinyl chloride monomer, approximately	20 g
Potassium persulfate ($K_2S_2O_8$) initiator	0.25 g
Tetraethylene glycol dimethacrylate (TEGDM) crosslinker	0.4 wt % of monomer

The PVC latex was filtered to remove any traces of coagulated polymer and was used as a seed latex for the second polymerization. New initiator (added in solution form as $0.25 \, g/10 \, ml \, H_2O$), plus TEGDM crosslinker (0.4 wt.% based on acrylonitrile plus butadiene) were stirred into the seed latex. The bottle was then sparged with nitrogen gas for about 5 min and the acrylonitrile monomer was weighed directly on top of the seed latex, 50 wt % based on PVC. Liquefied butadiene was introduced into the bottle directly on top of the floating acrylonitrile. The butadiene was allowed to boil off to the desired amount. The bottle was then capped and put into a 40°C water bath where it was tumbled for a minimum of 12 hr for a completely reacted product.

A portion of the finished latex (10–15 ml) was dried in open Petri dishes at room temperature to form cast films. The rest of the latex was coagulated (a small portion of saturated NaCl solution may be used), washed with water and 2-propanol, and finally molded at a pressure of about 5500 psi and at approximately 190°C to form suitable sheets.

Latex IPNs can be cast into integral films, provided that the glass transition temperature of the shell is at room temperature or below, and that the crosslink density is not too high. These criteria hold for the above syntheses. However, IPNs synthesized in the form of suspension-sized particles[29] cannot be either cast or molded properly. Apparently, the reason lies in the reduced surface area of the latter.

5.4.3. Latex Semi-IPNs

5.4.3.1. ABS Plastics

It is a matter of significant interest that many commercial ABS plastics made via an emulsion polymerization route are actually semi-I IPNs.[30,31] A typical recipe for polymer I is as follows:

Water	200 parts
Butadiene	100 parts
Divinyl benzene	2 parts
Sodium stearate	5 parts
Tert-dodecyl mercaptan	0.4 parts
Potassium persulfate	0.3 parts

The polymerization is carried out at 50°C, taking the usual precautions for removing dissolved oxygen, etc. The last several percent of butadiene must be removed. The grafting stage is also reacted at 50°C, after the following additions:

Water (including H_2O from the above)	300 parts
Polybutadiene (in latex form)	20 parts
Styrene	62 parts
Acrylonitrile	18 parts
Tert-dodecyl mercaptan	0.1 parts
Sodium stearate	0.5 parts
Potassium persulfate	0.5 parts

When the reaction is complete (100% conversion), 1% of a phenolic antioxidant is added, and the latex coagulated.

The points of interest refer to the polymerization of the seed latex. Since butadiene is a difunctional monomer, polymerization to a conversion of more than about 70% results in a crosslinked polymer. Since the reaction is relatively poorly controlled, the reproducibility of the recipe is improved with the addition of both a crosslinker and a chain-transfer agent. While some recipes yield a linear polymer I, apparently the preferred route utilizes a crosslinked elastomer because of improved dimensional stability of the elastomer through the processing steps.

5.4.3.2. Acrylic Compositions

In a broad study to evaluate the composite properties of hetero-geneous-latex-based materials, Dickie et al.[32] synthesized a series of latex semi-IPNs. A two-staged emulsion polymerization procedure was employed that yielded relatively uniform populations of heterogeneous latex particles (HLP). The glassy component was prepared from methyl methacrylate, and the rubbery component used a mixture comprising 95 mol % butyl acrylate (BA) and 5 mol % 1,3-butylene dimethacrylate (BPMA). Heterogeneous latex particles in which the rubbery component was polymerized first were referred to by Dickie et al. as HLP1; particles for which the order of polymerization was reversed were referred to as HLP2.

For the two-stage incremental addition polymerizations, a seed latex was prepared with approximately 2 parts surfactant, 50 parts monomer, 2 parts initiator, and 700 parts water. Additional monomer was added step-wise at 48–50°C. Upon completion of the first-stage polymerization, 50–500 parts of the second monomer charge were added incrementally at the same temperature.

5.5. SIMULTANEOUS INTERPENETRATING NETWORKS (SINs)

A SIN synthesis refers to mixing of monomers, prepolymers, linear polymers, crosslinkers, etc., of both polymers to form a homogeneous fluid. Both components are then simultaneously polymerized by independent, noninterfering reactions.* Several subcases have been explored: (a) the simultaneous gelation of both polymers,[33–35] (b) a sequential poly-merization of the prepolymer mix,[36–46] and (c) the introduction of a greater or lesser number of graft sites between the two polymers.[47–54] A major advantage of SINs over sequential IPNs relates to ease of processing. For example, the mix can be prepolymerized until just short of the gel point, followed by pumping into a mold or die, with continued polymerization.[11–14] For reasons of this ilk, more attention has been devoted to SIN-type IPNs than any other type. Typical reaction schemes will now be given for each of the three cases mentioned above.

---

* Topological designation: $[C_1 O_1 C_2]$

5.5.1. Simultaneous Gelation

The main idea was to polymerize two networks, in SIN form, such that both networks reached the Flory gelation point more or less simultaneously. In order to insure independence of the reactions, step and chain polymerization pairs were selected. One pair chosen for intensive study was based on an epoxy/acrylic mix.[34] The step component was made by reacting a low-molecular-weight epoxy resin, Epon 828, with 31.1 parts of phthalic anhydride per hundred parts of resin. The chain component was made by adding a mix of n-butyl acrylate with 1.6% diethylene glycol dimethacrylate (DEGDM) with 2% isoprene included for staining with osmium tetroxide, and di-t-butyl peroxide initiator at varying concentrations (based on nBA) from 0.16% to 0.40%. This forms a plastic/rubber composition, of course. Varying the peroxide concentration changes the rate of the acrylic–rubber polymerization, allowing it to be relatively slower, approximately simultaneous, or faster than the epoxy polymerization. See Figure 5.2 for a summary of chemical components and reaction conditions.

Conditions approximating simultaneous gelation were established in studies on both homopolymers. For the SINs, the samples were made by heating the Epon 828 to 120°C and then adding the phthalic anhydride with stirring. After about 20 min, when the phthalic anhydride was dissolved, the n-butyl acrylate mix was added and the reaction flask was purged with dry nitrogen for 5–10 min and then sealed. The reaction was continued at 120°C with stirring for about 21–31 hr, until it became viscous enough to prevent the lighter rubber component from coalescing and rising to the surface. The mix was then poured into heated glass plate molds. The reaction was run for at least three days at 120°C and then heated to 150°C for an additional day to insure full curing of the epoxy. It should be noted that the half-life of the di-t-butyl peroxide at 120°C is approximately 20 hr.

Surprisingly, truly simultaneous gelation resulted in a minimum in the physical properties (see Section 7.3.3). While truly simultaneous gelation

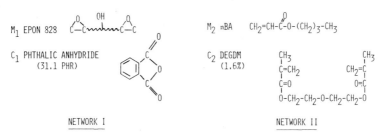

Figure 5.2. Principal components employed in the epoxy/acrylic SIN syntheses.[34] Initiator: Di-t-butyl peroxide, 0.16, 0.24, 0.32, 0.40%. Reaction conditions: T = 120°C, 3 days: T = 150°C, 1 day.

may please the intellect, apparently much stronger materials were formed when one or the other of the two reactions was run slightly faster.[34,35]

5.5.2. Sequential Polymerization of a Prepolymer Mix

In a polymerization of this type, one or both monomers are first polymerized to form a prepolymer. The two prepolymers are mutually dissolved. Then they are reacted to the network stage more or less sequentially.

5.5.2.1. Polyurethane Compositions

Kim et al.[39] investigated a system where the principal components were a polyurethane and poly(methyl methacrylate). The synthesis may be briefly outlined as follows:

Materials. The raw materials used are described in Table 5.5.[37] Poly(caprolactone) glycol with average molecular weight of 1978 (Niax D560), 1,4-butanediol (1,4-BD), and trimethylolpropane (TMP) were dried at 60°C for 5 hr under a vacuum of 2 mm. Methyl methacrylate (MMA) was washed with 5% aqueous potassium hydroxide solution, followed by washing with distilled water, dried over Linde 4A molecular sieves, and distilled at 40°C under a vacuum of 2 mm to remove the stabilizer. Trimethylolpropane trimethacrylate (TMPTMA) was also distilled at 40°C under 2 mm vacuum. 4,4'-Diphenylmethane diisocyanate (MDI) and benzoyl peroxide were used without further purification.

Preparation: Polyurethane (PU). The isocyanate-terminated polyurethane prepolymer was prepared by reacting two equivalent weights of 4,4'-diphenylmethane diisocyanate (MDI) with one equivalent weight of poly(caprolactone) glycol (Niax D-560) at 60°C. The prepolymer was stored under vacuum (for not more than two days) because of its susceptibility to

Table 5.5. Polyurethane SIN Materials[37]

Designation	Description	Source
Niax D-560	Poly(caprolactone) glycol mol wt = 1978, OH No. = 56.7	Union Carbide Corp.
TMP	Trimethylolpropane	Celanese Chem. Corp.
1,4-BD	1,4-Butanediol	GAF Corp.
MDI	4,4'-Diphenylmethane diisocyanate	Mobay Chem. Co.
MMA	Methyl methacrylate	Fisher Sci. Co.
TMPTMA	Trimethylolpropane trimethacrylate	Polyscience, Inc.
BPO	Benzoyl peroxide	Fisher Sci. Co.

moisture. One equivalent weight of the PU prepolymer was heated to 80°C and then homogeneously mixed with one equivalent weight of a 1,4-butanediol (1,4-BD) and trimethylolpropane (TMP) mixture (4:1 equivalent ratio) for 5 min using a high-torque stirrer. The air entrapped during mixing was removed by applying a vacuum for 5–10 min. The mixture was cast in a closed stainless steel mold (with polypropylene lining for easy demolding) at 80°C for 16 hr and 110°C for 4 hr on a platen press under 350 psi pressure.

   Poly(methyl methacrylate) (*PMMA*). A weight of 118.3 g of distilled MMA monomer, 1.7 g of distilled TMPTMA, and 1.2 g of benzoyl peroxide were reacted in a resin kettle. The mixture was stirred until the benzoyl peroxide was dissolved. The progress of the reaction was followed by placing a drop of mixture in isopropyl alcohol where the PMMA precipitated. The reaction was continued until 10–15% conversion and stopped by rapid cooling. The reaction mixture was cast and cured at 80°C for 16 hr and 110°C for 4 hr under 350 psi pressure.

   IPNs. One equivalent weight of the PU prepolymer was heated to 80°C and then homogeneously mixed with one equivalent weight of the 1,4-BD–TMP mixture (4:1 equivalent ratio) for 5 min. Then the MMA–TMPTMA prepolymer mixture was added in varying weight ratios and homogeneously mixed for 3 min using a high-torque stirrer. The air entrapped during mixing was removed by applying a vacuum for 30 sec. The mixture was then cast in the same manner as described above.

   Several semi-SINs and chemical blends* were also prepared:

   Linear PU was prepared in the same manner as for the polyurethane preparation, except that one equivalent weight of PU prepolymer was reacted with one equivalent weight of 1,4-BD instead of 1,4-BD–TMP mixture. Linear PMMA was prepared in the same manner as for the PMMA preparation except that the crosslinking agent (TMPTMA) was omitted.

   Simultaneous chemical blends were prepared for 75%–25% and 50%–50% PU–PMMA compositions in the same manner as the SIN preparation except that the 1,4-BD–TMP mixture was replaced by pure 1,4-BD in the PU network and the TMPTMA was omitted in the PMMA network.

   Two semi-SINs were prepared for the 75% PU–25% PMMA composition. They were prepared in the same manner as the IPN preparation except that one or the other of the component networks was made linear by the procedure described above.

---

* Kim et al.[37] refer to pseudo-IPNs and linear blends in their notation. The term chemical blends refers to the polymerization of one (or both) monomer(s) in the presence of the other, to distinguish from the term mechanical blends, obtained by mixing previously formed polymers, and is actually a class of graft copolymer.

A number of samples were prepared. Their designations are given in Table 5.6.[37] The symbols U and M refer to urethane and methacrylate, respectively, and C and L refer to crosslinked and linear, respectively.

5.5.2.2. Castor Oil/Polystyrene

Castor oil–sebacic acid polyesters and polyurethanes were prepared in SIN form with polystyrene by Devia *et al.*[42–46] Since sebacic acid is commercially derived from castor oil, a 100% castor oil elastomer was prepared. The synthesis is as follows.

Three different crosslinking agents for the castor oil were used: (a) sebacic acid or derivatives to form a castor oil polyester network (COPEN), (b) 2,4-tolylene diisocyanate (TDI) to form a castor oil polyurethane network (COPUN), and (c) sebacic acid plus 2,4-TDI to form castor oil poly(ester-urethane) network (COPEUN). In each case, the three hydroxyl groups on the castor oil molecule were reacted [structure (5.1)]. The synthesis procedure started with the preparation of the corresponding prepolymers: (a) castor oil polyester prepolymer, COPEP1, the resultant product of the reaction of one castor oil equivalent weight with one sebacic acid equivalent weight until the acid value fell to 33; (b) COPEP4, an extended chain polyol obtained by completely reacting 0.6 acid equivalent weights with one castor oil equivalent weight; (c) castor oil polyurethane

Table 5.6. Polyurethane SIN Samples[37]

Sample code	Composition and crosslink level	Type of material
UC100	Crosslinked PU 100%	Homopolymer
UL100	Linear PU 100%	Homopolymer
MC100	Crosslinked PMMA 100%	Homopolymer
ML100	Linear PMMA 100%	Homopolymer
UC85MC15	Crosslinked PU 85%, crosslinked PMMA 15%	SIN
UC75MC25	Crosslinked PU 75%, crosslinked PMMA 25%	SIN
UC60MC40	Crosslinked PU 60%, crosslinked PMMA 40%	SIN
UC50MC50	Crosslinked PU 50%, crosslinked PMMA 50%	SIN
UC40MC60	Crosslinked PU 40%, crosslinked PMMA 60%	SIN
UC25MC75	Crosslinked PU 25%, crosslinked PMMA 75%	SIN
UC15MC85	Crosslinked PU 15%, crosslinked PMMA 85%	SIN
UC75ML25	Crosslinked PU 75%, linear PMMA 25%	semi-SIN
UL75MC25	Linear PU 75%, crosslinked PMMA 25%	semi-SIN
UL75ML25	Linear PU 75%, linear PMMA 25%	Chemical blend
UL50ML50	Linear PU 50%, linear PMMA 50%	Chemical blend

prepolymer, COPUP1, an isocyanate-terminated prepolymer formed from the reaction of 2.2 equivalent weights of 2,4-TDI with one equivalent weight of castor oil.

The styrene mixture was prepared by dissolving proportionally 0.4 g of benzoyl peroxide and 1 ml of commercial divinylbenzene solution (55%) in 99 ml of freshly distilled styrene monomer. This was polymerized to form a polystyrene network, PSN. SINs containing 10 and 40% castor oil elastomer were studied.

The reaction scheme worked out for the castor oil–sebacic acid/polystyrene SIN will serve as a prototype example for the other materials.

Castor oil and sebacic acid are reacted at 180–200°C until the mixture approaches gelation, so a branched prepolymer having an equal number of both functional groups (COPEP1) is obtained (Figure 5.3, upper left). The reaction is then stopped by cooling the prepolymer to 80°C. The styrene-DVB mixture is prepared at room temperature and charged to the reactor containing the polyester prepolymer, where mixing takes place (Figure 5.3, lower left). This yields a mutual solution of all components required for the formation of both networks. The temperature is then raised to 80°C in order to initiate the styrene polymerization. (The polyester reaction rate is nil at this temperature.) In polymerizing the styrene component within the polyester prepolymer mixture, the first polystyrene produced remains dissolved until some critical concentration is reached, followed by phase separation.

The solution is transformed to an oil-in-oil emulsion in which a polystyrene solution forms the disperse phase and the elastomer polyester component solution the continuous phase. The point of phase separation is observed experimentally by the onset of turbidity, owing to the Tyndall effect. The conversion required for phase separation to occur depends basically on the solubility of the polystyrene chains in the castor oil prepolymer.

As polymerization proceeds, the total volume of polystyrene polymer particles increases rapidly at the expense of the styrene monomer from the solution. What happens next depends on several factors, mainly composition and stirring. It was found that for SIN formulations having an elastomer content greater than about 15%, no further changes occur and the elastomer material will remain the continuous phase, regardless of the extent of agitation.

However, for SINs having up to 10–15% elastomer content, it was found that stirring induces significant changes in the morphology of the mixture. (If stirring is not provided, the polystyrene polymer particles will sink and coalesce giving rise to a two-layered system.)

Anticipating the material presented in Section 6.3, the reacting mass must be stirred until phase inversion is complete. However, the mass must be

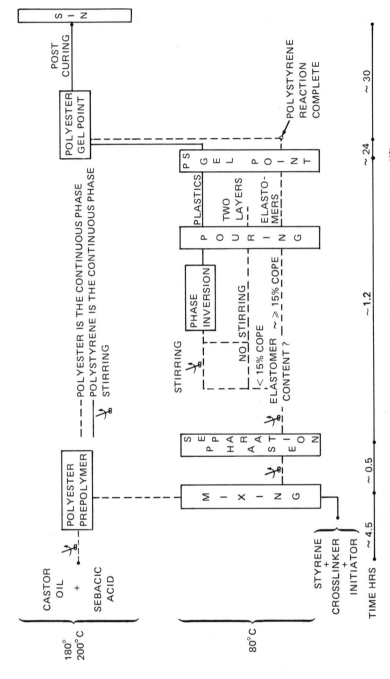

Figure 5.3. A process scheme for the synthesis of castor oil polyester SINs.[43]

transferred to the mold before gelation of the polystyrene takes place. Gelation is indicated by an increase in the second normal stress difference, i.e., the product climbs the stirrer. Products cast or transferred to the mold after this point will exhibit poor mechanical properties because they undergo a second phase inversion. Thus a window exists for the pouring time, bounded on the one side by phase inversion, and by gelation on the other.

5.5.2.3. Butyl Rubber SINs

Ozerkovskii et al.[56] studied a series of solution blends, semi-SINs, and SINs. The most interesting was the last, prepared by simultaneously synthesizing network II and crosslinking a preformed polymer I. Polymer I was butyl rubber, BR, and polymer II was poly(methyl acrylate), PMA, poly(butyl acrylate), PBA, or poly(nonyl acrylate), PNA (see Table 5.7[56]).

The SIN of PNA and BR was prepared as follows. Butyl rubber was dissolved in nonyl acrylate. Dinitrobenzene or trinitrobenzene was employed as a vulcanizing agent for the BR, which contained about 1% double bonds. The polymerization of the NA was initiated by benzoyl peroxide. A crosslinked network of PNA is formed by chain transfer to the α-hydrogen.

The concentrations of PNA and BR in the gel and sol fractions were determined by ir spectroscopy. In the absence of vulcanizing agents, the gel

Table 5.7. Calculated Surface Tension of Polymers (γ_i) and the Interfacial Tension (γ_{iBR}) in BR–polyacrylate SINs[a(56)]

Polymer	Structural formula of unit	γ_i, dyn/cm	γ_{iBR}, dyn/cm
Butyl rubber (BR)	$\begin{array}{c} CH_3 \\ \mid \\ -CH_2-C- \\ \mid \\ CH_3 \end{array}$	27	0
Poly(methyl acrylate) (PBR)	$\begin{array}{c} -CH_2-CH- \\ \mid \\ O=C-O-CH_3 \end{array}$	53	4.41
Poly(butyl acrylate) (PBA)	$\begin{array}{c} -CH_2-CH- \\ \mid \\ O=C-O-C_4H_9 \end{array}$	46	2.56
Poly(nonyl acrylate) (PNA)	$\begin{array}{c} -CH_2-CH- \\ \mid \\ O=C-O-C_9H_{19} \end{array}$	41	1.44

[a] $\gamma_i = (P/V)^4$ (P = parachor, V = molar volume); $\gamma_{12} = (\gamma_1^{0.5} - \gamma_2^{0.5})^2$.

fraction consisted almost entirely of PNA. The sol fraction consisted entirely of BR. Hence, under these conditions grafting of the growing PNA chains to the BR does not occur. In the presence of dinitrobenzene or trinitrobenzene, vulcanization of the BR occurs simultaneously with the formation of a PNA network, and a composite containing interpenetrating networks of the two components is formed.

5.5.2.4. Polyurethane/Poly(methyl methacrylate) Semi-SINs

Allen *et al.*[57] synthesized a series of semi-SINs based on polyurethane elastomers and methyl methacrylate. The general scheme was as follows*:

$$\text{elastomer precursor + vinyl monomer} \xrightarrow{\text{urethane catalyst}}$$

$$\text{gel-containing vinyl monomer} \xrightarrow{\text{vinyl initiator}} \text{composite} \qquad (5.2)$$

The reaction was carried out as a one-shot, two-step procedure. At room temperature, dibutyltin-dilaurate (DBTL) was the urethane catalyst. Trifunctional polyols served to crosslink the polyurethane. The temperature was then raised to initiate the MMA polymerization. An example of the Allen *et al.*[57] synthesis is as follows.

To 64 g methyl methacrylate (MMA), 0.16 g azobisisobutyronitrile (AIBN), 1.93 g MDI, 7.03 g poly(oxypropylene) glycol (PPG-2000), 7.03 g poly(oxypropylene) adduct of glycerol (OPG-3000), and 0.26 g DBTL were added and mixed in a conical flask and allowed to stand at room temperature. For the above compositions [80:20 MMA/PU, molar ratio NCO/OH = 1.1:1], gelation took about 4–5 hr. Prior to gelation the sample was quickly degassed, poured into a mould, and allowed to gel. After initial mixing, no stirring was undertaken. After gelation, the casting was cured at 50°C for 15 hr, 90°C for 1 hr, and 115–120°C for a further 2 hr.

5.5.2.5. Topotactic Polymerizations

Belonovskaya *et al.*[58] carried out a novel series of syntheses of materials based on the concept of a semi-SIN. Belonovskaya *et al.* called their material an interpenetrating polymer system, IPS. They designate such a reaction as a topotactic "replica," or matrix polymerization. Such a synthesis yields a compatible, high-temperature polymer combination.

---

* Note that it is better to write that the polyurethane gel contains the vinyl monomer, rather than is swollen with it, because the urethane chains are in a relaxed conformation.

Monomer I (M_1) was either 2,4-tolylene-diisocyanate (TDI) or hexamethylene diisocyanate (HDI). Monomer II (M_2) was one of the following polar monomers: ethylene and propylene sulfide (ES and PS), methyl methacrylate (MMA), methyl acrylate (MA), acrylonitrile (AN), acrolein (ACR), and methacrolein (MACR). Initators included triethylamine (TEA), tetraethylene diamine (TMEDA), trimethylenediamine (DABCO), benzoyl peroxide, and dinitroazobutyric acid.

A two-stage polymerization was carried out under argon. Owing to the high reactivity of the isocyanates, they polymerize first according to the following reaction scheme:

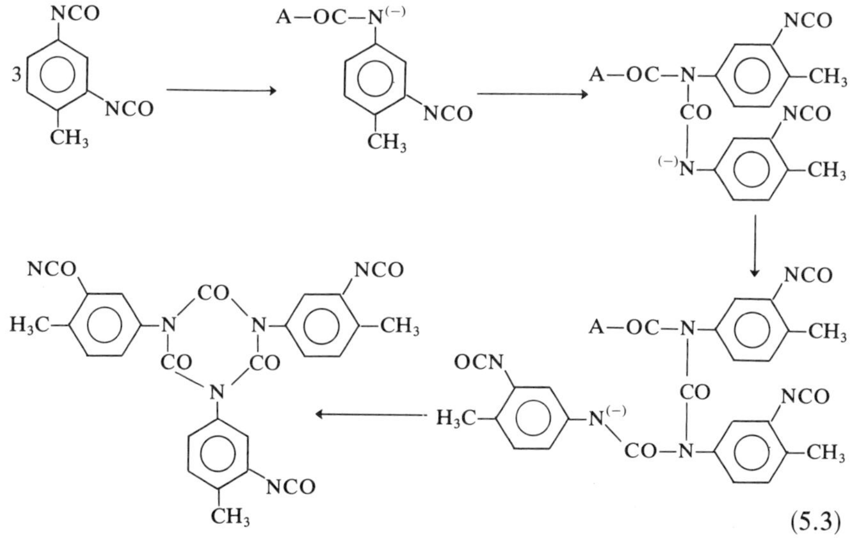

(5.3)

where A is a nucleophilic initiating agent. (Interestingly, the final product shows an isocyanate-containing isocyanurate.) It should be noted that the polyisocyanates made by Belonovskaya are not polyurethanes, but rather extremely tight, highly polar networks. The structure of poly-alkylenediisocyanates is more complex. It was suggested that the HDI polymer contains different structures:

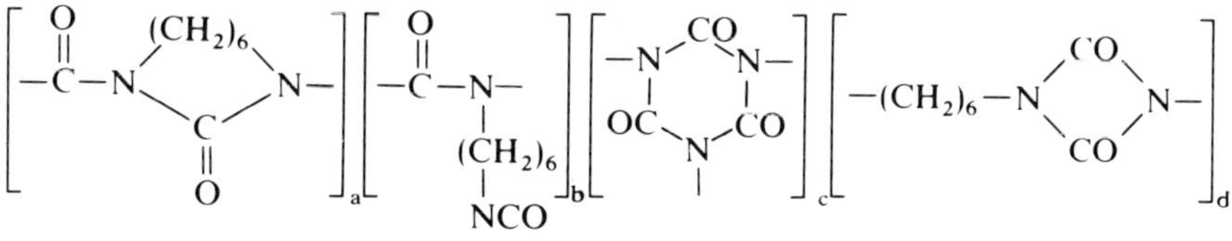

(5.3')

where a, b, c, and d refer to the number of moles of each mer in the polymer.

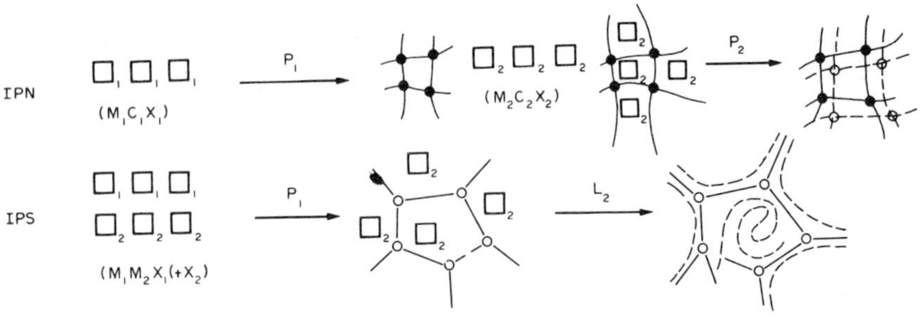

Figure 5.4. A reaction scheme differentiating an IPN from an IPS. The IPS is a variation on the semi-SIN idea, where the two monomers are mixed, followed by a sequential polymerization. The first polymer is crosslinked, and the second linear.[58] M_1, M_2—Addition of monomers; C_1, C_2—addition of crosslinking agents; X_1, X_2—addition of catalysts; P_1, P_2—formation of polymer network; L_2—formation of linear polymer; $\overset{\backprime}{\mid}\overset{\prime}{}$ —triisocyanurate ring.

In the IPS system, M_1 and M_2 were added and subjected to vigorous stirring until gel-formation was achieved. Then the product was allowed to stand up and harden, followed by a stepwise heating of the samples. The agent A^- initiates the M_2 monomer, after the M_1 reaction is complete.

The reaction scheme, differentiating an IPS from an IPN, is given in Figure 5.4.[58] The group theory notation is derived from papers by Sperling[59,60] (see Chapter 3). The structure in Figure 5.4 suggests a particular morphology. Belonovskaya et al.[58] state that the polymerization of M_2 differs greatly from the corresponding free polymerization in bulk or in solution. The polar monomer, M_2, is fixed in the pores of the rigid polyisocyanate network, and its orientation is due to polar interactions with the triisocyanate rings, forming a "physical" network.*

5.5.3. Grafted SINs

In any IPN or SIN polymerizations, some grafting is inevitable. As a first approximation, however, grafting may be neglected if the following two criteria hold: (1) the concentration of crosslinks within each network significantly exceeds the concentration of graft sites, and (2) the morphology and behavior are relatively unaffected by the actual extent of grafting.†

In most real SINs, unfortunately, neither criterion may be established with certainty. Since the materials described in Sections 5.5.1. and 5.5.2

* An analogy might be drawn to some inorganic systems, such as cuprous oxides, Cu_2O, which form the two interpenetrating frameworks (see A. F. Wells.[61])

† It must be emphasized that modest levels of grafting improve the interfacial bonding, and hence the mechanical properties profit.

were distinguished by well-defined two-phased morphologies and noninterfering chemical reactions they were arbitrarily set apart from the compositions to be discussed below.

5.5.3.1. Grafted Urethane SINs

An example of a SIN synthesis now thought to contain significant grafting was developed by Frisch *et al.*[50]:

Materials. The materials used are listed in Table 5.8. All polyols were dried at 80°C for 10 hr under a vacuum of 0.1 Torr. The solvents used were reagent grade and dried over molecular sieves.

Preparation: Polyacrylate (PA). The polymer used was a 50% solution in xylene and cellosolve acetate, and consisted of random copolymer of butyl

Table 5.8. Compatible Urethane SIN Materials[50]

Designation	Description	Source
Polymeg 660	Poly(1,4-oxybutylene) glycol [poly(tetramethylene) glycol] MW = 661, hydroxyl number 169.8	Quaker Oats Co.
Polymeg 1000	Poly(1,4-oxybutylene) glycol [poly(tetramethylene) glycol] MW = 1004, hydroxyl number 111.8	Quaker Oats Co.
TMP	Trimethylolpropane	Celanese Chem. Corp.
2-But. Ox.	2-Butanone oxime	Matheson Coleman & Bell
$H_{12}MDI$	4,4'-Methylene bis(cyclohexyl isocyanate)	Allied Chem. Co.
MDI	4,4'-Diphenylmethane diisocyanate	Mobay Chem. Co.
XDI	Xylylene diisocyanate; 70/30 mixture of *meta* and *para* isomers; NCO = 94.1	Takeda Chem. Co.
T-12	Dibutyltin dilaurate	M & T Chem. Co.
T-9	Stannous octoate	M & T Chem. Co.
Acrylic 342–CD 725	Random copolymer of butyl acrylate, methacrylic acid, styrene and hydroxyethyl methacrylate; 50% solution in xylene: Cellosolve acetate (1:1); hydroxyl number 60.0; acid number 13.5	Inmont Corp.
Melamine RU 522	Butylated melamine formaldehyde resin; 60% solution in xylene: Cellusolve acetate (1:1)	Inmont Corp.
Silicone L-522	Polydimethylsiloxane-polyoxyalkylene copolymer	Union Carbide Corp.
CAB	Celiolose acetate butyrate EAB-381-2; ASTM viscosity 15	Eastman Chem. Corp.

acrylate, styrene, and hydroxethyl methacrylate. It also contained a small amount ($<1\%$) of methacrylic acid, which was present as a catalyst for the melamine cure. It was crosslinked by reaction of the pendant hydroxyl groups with a butylated melamine formaldehyde resin (in 60% solution in xylene and cellosolve acetate).

Polyurethanes (PU). A number of different polyurethanes were synthesized in order to better determine structure–property relationships in the SINs. Five different isocyanate-terminated urethane prepolymers (all polyether based) were prepared: poly(tetramethylene) glycol, MW = 661 (PM 660) + 4,4'-diphenylmethane diisocyanate (MDI); PM 660 + xylene diisocyanate (XDI); and trimethylolpropane (TMP) + H_{12}MDI. A resin kettle equipped with a nitrogen inlet, stirrer, thermometer, and reflux condenser was charged with a 50% solution of two equivalent weights of isocyanate in a 1:1 mixture of Cellusolve acetate and xylene. To this was added slowly with stirring a 50% solution of one equivalent weight of polyol in the above solvent mixture. The reactions were carried out under nitrogen at 60°C (for MDI) and 80°C (for XDI and H_{12}MDI) until the theoretical isocyanate contents (as determined by the di-*n*-butylamine method) were reached.

Blocking. A 50% solution (in the above solvent mixture) of a slight equivalent excess of 2-butanone oxime and 0.2 wt % dibutyltin dilaurate (T-12) were added to the prepolymer solutions in a three-necked flask equipped with a stirrer, reflux condenser, thermometer, and nitrogen inlet. The reaction was carried out under nitrogen at 80°C until the isocyanate content reached zero (complete blocking).

Chain Extension and Curing. An equivalent weight of TMP (in a 50% solution as above), 0.1 wt % stannous octoate (T-9), a 1 wt % flow agent composed of a 1:1 mixture of cellulose acetate butyrate, and a poly-dimethylsiloxane–polyoxyalkylene copolymer (L-522) were added to the prepolymer solutions. Films were cast and cured at 150°C for 4 hr. At this temperature, deblocking occurs followed by chain extension and cross-linking.

SINs. The polyurethane solutions (containing crosslinking agents and catalysts) were thoroughly mixed with the polyacrylate–melamine solution. Combinations with 25% PU, 50% PU, and 75% PU were made. Films were cast and cured as above. They were all very clear. Thus five SINs were produced: SIN 1, PA–PU (PM660-H_{12}-MDI-TMP); SIN 2, PA–PU (PM1000-H_{12}MDI-TMP); SIN 3, PA–PU (PM660-MDI-TMP); SIN 4, PA–PU (PM 660-XDI-TMP); SIN 5, PA–PU (TMP-H_{12}MDI-TMP).

The authors recognized the grafting reaction between the isocyanate and the pendant hydroxyls on the polyacrylate, and blocked the isocyanate with 2-butanone oxime. However, all of these materials apparently formed one phase, which raised speculations regarding other possible reactions. In a later paper, Klempner et al.[55] discussed the possibility of reaction between the butylated melamine–formaldehyde resin and the hydroxy-terminated chain extender for the polyurethane. When the experiment was essentially repeated, but using ethylene glycol dimethacrylate as the acrylic cross-linker,[55] a two-phased structure was obtained.

The material employed in this latter study[55] was a copolymer of eight parts of n-butyl methacrylate, one part of ethyl methacrylate, and one part of styrene. The reaction was carried out at 80°C until a prepolymer of syrupy consistency was obtained. As used for SIN formation, various amounts of ethylene glycol dimethacrylate (EGDMA) were added, and mixed with the polyurethane prepolymer. In passing, it should be noted that the portion of acrylic polymerized during the prepolymerization step, sans EGDMA, forms a linear polymer, and apparently does not take part in the network formation.

5.5.3.2. Grafted Epoxy SINs

In the above, the grafting was caused primarily by side reactions. Deliberate grafting can be induced by several means: (1) radiation treatment after polymerization, (2) condensation reactions between the two polymers, or (3) introduction of a monomer reactive with both networks during polymerization. This last one will be considered here.

In order to examine the effect of grafting on SIN formation, Scarito and Sperling[54] essentially repeated the experiments by Touhsaent et al.[34,35] on epoxy/acrylic combinations (described above), but added various levels of glycidyl methacrylate. The glycidyl methacrylate reacts with both materials during SIN formation, yielding well-defined graft sites. At a level of 3% of glycidyl methacrylate, only one glass transition was found. Thus, only relatively small amounts of internetwork grafts are required to profoundly change the physical behavior of the final product.

5.5.3.3. Further Polyurethane SIN Syntheses

Frisch et al. also prepared several other SIN-type IPNs which exhibited one phase behavior. For that reason, the following syntheses[48] are included in the present discussion, even though it is not known whether grafting per se has caused the apparent compatibility.

The materials used and their descriptions are listed in Table 5.9. Three urethane networks, two polyether based (PU-1,2) and the one polyester based (PU-3), were synthesized.

Table 5.9. Materials for Compatible Polyurethane SINs[48]

Designation	Description	Source
Polyester P-373	Unsaturated polyester dipropylene glycol maleate	Marco Division, W. R. Grace & Co.
Epon 828	Epoxy resin composed of bisphenol A and epichlorohydrin	Shell Chemical Co.
TDI	Tolylene diisocyanate; 80/20 mixture of 2,4 and 2,6 isomers NCO:87.0	BASF Wyandotte Corp.
$H_{12}MDI$	4,4'-methylene bis (cyclohexyl isocyanate)	Allied Chem. Co.
Pluracol TP-440	Poly (xypropylene) adduct of trimethylolpropane (urethane grade) MW = 420, hydroxyl No. = 322	BASF Wyandotte Corp.
Elastanol JX2057	Hydroxy-terminated polyester of 1,4-butanediol and adipic acid; MW = 420, hydroxyl No. = 55.1	North American Urethanes, BASF Wyandotte Corp.
DMP-30	2,4,6-tris (Dimethylaminomethyl phenol	Rohm & Haas Co.
Polymeg 660	Poly (1,4-oxybutylene) glycol [poly(tetramethylene) glycol]; MW = 661, hydroxyl No. = 187	Quaker Oats Co.
Benzoyl Peroxide		
T-9	Stannous octoate	M&T Chem., Inc.
T-12	Dibutyltin dilaurate	M&T Chem., Inc.
Maleic Anhydride		Monsanto Co.
Adipic Acid		Monsanto Co.
DEG	Diethylene glycol	Dow Chemical Co.
DPG	Dipropylene glycol	Dow Chemical Co.
CAB	Cellulose acetate butyrate EAB-381-2 ASTM Viscosity 15	Eastman Chemical Products, Inc.
Silicone L522	Poly (dimethyl siloxane)-poly (oxyalkylene) copolymer	Union Carbide Corp.
TMP	Trimethylolpropane	Celanese Chem. Co.
Styrene	Monomer	Dow Chemical Co.

PU-1. This polyether prepolymer (NCO/OH = 2) was prepared at 80°C under nitrogen. 1.5 equivalent weights of poly (tetramethylene glycol), MW = 661 (PM 660), were slowly added with stirring to three equivalent weights of tolylene diisocyanate (TDI) in a resin kettle. The reaction was allowed to proceed until the theoretical isocyanate content (determined by the di-*n*-butylamine method) was reached (2 hr). An equivalent weight of a poly(oxypropylene) adduct of trimethylolpropane (MW = 420) (TP-440) was added to the prepolymer, as well as 0.05 parts of stannous octoate (T-9) (catalyst), and the mixture cured in a stainless steel mold at 100°C for 16 hr.

PU-3. The prepolymer was also made at 80°C under nitrogen. Two equivalent weights of TP-440 were added slowly with stirring to four equivalent weights of TDI. The reaction was carried out until the theoretical isocyanate content was reached. An equivalent weight of a hydroxyl terminated polyester, consisting of 1,4-butanediol and adipic acid (MW = 2036) (Elastanol JX2057) was added to the prepolymer and the mixture diluted to 50% solids with cellosolve acetate. The mixture was reacted at 70°C for 1 hr under nitrogen. To this solution was added 1 wt % flow agent and 0.2 wt % T-9 catalyst. Films were then cast on glass and cured at 85°C for 16 hr and 130°C for 2 hr.

Polyesters. Two unsaturated polyesters were employed, one of which was highly unsaturated (dipropylene glycol maleate) (P-373), while the other was more flexible, containing adipic acid in addition to maleic anhydride. Both were crosslinked by means of styrene monomer.

P-373. To 100 g of the unsaturated prepolymer were added 30 g of styrene and 1.3 g benzoyl peroxide. The mixture was stirred and castings made between glass plates sealed with rubber gaskets (to prevent monomer evaporation) and cured overnight at 85°C.

PE 1. A mixture of 0.2 mol maleic anhydride, 0.8 mol adipic acid, 0.2 mol dipropylene glycol, and 0.9 mol diethylene glycol was reacted at 150°C under nitrogen for 24 hr. The resultant prepolymer was cured with styrene and films made as above.

Epoxy. A 50% solution of Epon 828 (bisphenol A–epichlorohydrin resin) in Cellosolve acetate was made. To this was added 1 wt % flow agent (see above) and 0.5 wt % 2,4,6-tris(dimethylaminomethyl) phenol (DMP-30). Films were cast on glass as above and cured at 85°C for 16 hr and 130°C for 2 hr.

SINs. Five different polymer combinations were made, two in bulk and three in solution. They were made by combining the linear prepolymers, crosslinking agents and catalysts, casting films or molding the mixtures, and curing them *in situ*, thereby forming SINs.

SIN 1. Equal weights of the polyurethane solution PU-3 (containing flow agent and catalyst) and the epoxy solution (also containing flow agent and catalyst) were mixed. Films were cast on glass with a doctor blade and cured at 85°C for 16 hr and 130°C for 2 hr. Completely clear films resulted.

SIN 2. This SIN was the same as SIN 1 but was composed of 25% PU-3 and 75% epoxy. The resulting films were perfectly clear.

SIN 3. Equal weights of polyester P-373 (including styrene and benzoyl peroxide) and polyurethane PU-1 (including TP-440 and T-9) were mixed. No solvent was added because the combination had a low enough viscosity, owing to the presence of monomeric styrene, to allow mixing. The mixture was cast between glass plates and cured at 110°C for 16 hr. The specimens were all perfectly clear.

SIN 4. This was made in the same way as SIN 3; however, polyester PE-1 was used.

For brevity, the synthesis of a polyacrylate, very similar to that discussed above, has been omitted, as well as the corresponding polyurethane and SIN.

It is noted that the polyester synthesis itself constitutes an AB-crosslinked copolymer (see Section 5.6), and has the topological designation $P_1 O_C P_2$, where P_1 = polyester and P_2 = polystyrene. If subscript 3 indicates the polyurethane, the SIN may be designated $[(P_1 O_C P_2) O_I C_3]$, where the brackets indicate simultaneous reactions (see Chapter 3).

Meyer and Mehrenberger[62] also prepared grafted SINs based on an unsaturated polyester-styrene/polyurethane system (see SIN 3, above).[48] The synthetic details employed by Meyer and Mehrenberger were as follows.

a. *Polyurethane.* A commercially available dipropylene glycol, PLURACOL P 410, marketed by Ugine Kuhlmann Co. was used. Just before use, it was dried in vacuum at 80°C. The isocyanate was Desmodur AP (Bayer Company). It was an adduct of tolylene diisocyanate and trimethylolpropane for which the remaining (three) isocyanate functions were blocked by phenol. Heating (150°C) decomposed the adduct into triisocyanate and phenol. Isocyanate content: 12%.

b. *Polyester.* A commercial product, UKAPON D 20, marketed by Ugine Kuhlmann Co. was used. It is an unsaturated ester containing 34.8 wt % styrene. Crosslinking occurs via styrene reacting with the polyester double bonds, yielding a rigid network. The reaction was catalyzed at room temperature by methyl ethyl ketone peroxide and may be accelerated by cobalt naphthenate. Acid number: 33.

The individual networks were synthesized as follows.[62]

a. *Polyurethane Network (PU).* A solution of Desmodur AP (50 parts) and Cellosolve acetate (50 parts) was prepared. 350 parts of this solution were mixed with 100 parts of diol, corresponding to NCO : OH = 1 : 1. This mixture was thoroughly agitated and poured into an aluminum mold. The mold was heated (in an oven) for 3 hr at 150°C. The resulting film was cured for another 16 hr at the same temperature. PU films were transparent and lightly red colored.

b. *Polyester Network (PE).* Reaction was very fast when both initiator and accelerator were utilized: at room temperature, the reaction was complete after 1 hr. Without the accelerator, a gel formed in 1 hr at 60°C; at

Table 5.10. Polyester-Styrene/Polyurethane
SINs[62]

SIN No.	PU (wt %)	PE (wt %)
1	7	93
2	22	78
3	40	60
4	60	40
5	85	15

150°C, heat-induced polymerization gave a gel within a few minutes, and a solid transparent yellow film in 30 min. Aluminum molds were also used. Experimental conditions (150°C) close to those for PU networks were retained for IPN preparation.

Meyer and Mehrenberger[62] synthesized their IPNs as follows.

A mixture of PE and PU solution (starting viscosity: about 500 cP) was agitated for 30 min poured into aluminum molds and heated at 150°C for 16 hr. Table 5.10 lists the composition of the SINs prepared by Meyer and Mehrenberger. They described their films as usually transparent, light red in color, and without any apparent heterogeneity. They also found a single glass transition for all compositions, although the temperature of the transition and its breadth behaved somewhat differently than might be expected for total compatibility.

Meyer and Mehrenberger found that viscosities of various mixtures of the components remained constant over several days, suggesting that little or no intersystem reaction occurred. The extensive compatibility of their system might be explained, however, by assuming sufficient grafting between the polyester (—OH on one end and —COOH on the other), and the polymerizing isocyanate-based urethane to increase compatibility. The level of such grafting might be fairly low.

5.6. AB-CROSSLINKED POLYMERS (ABCPs)

An ABCP contains two polymers which are grafted together in such a way as to form one monolithic network.* Two subclasses may be distinguished.

a. Polymer II is grafted to polymer I in a conterminous fashion, i.e., both ends of the polymer II chain are bound to different polymer I chains.

---

* Topological designation: $P_1 \, O_C \, P_2$.

b. The polymer II chains grow across polymer I chains, forming a series of tetrafunctional graft linkages.

Since an IPN contains two polymers each in separate network form, clearly an ABCP is not an IPN. However, it constitutes an interesting close relative. Further, if either or both polymers in an ABCP are crosslinked then a grafted semi-IPN or IPN is formed. Thus the materials sometimes look, topologically, like the grafted structures discussed in Section 5.5.

5.6.1. Poly(vinyl trichloroacetate)-Based ABCPs

Bamford and Eastmond synthesized well-defined ABCP structures,[63-67] as illustrated in Figure 5.5. In the synthesis of ABCP polymers, polymers carrying reactive halogen-containing groups are employed. Such groups are incorporated into the side chains of the polymers. Typical polymers used in the preparation of ABCPs contain on the order of 10^3 such groups per polymer molecule. Most of their studies have used poly(vinyl trichloroacetate) (PVTCA) as polymer I. The chlorines are activated by $Mo(CO)_6$ thermally at 80°C, or via photoinitiation of $Mn_2(CO)_{10}$ at 25°C. Polymers which have been used in this connection besides (PVTCA) include poly(vinyl bromoacetate), cellulose acetate containing a proportion of trichloracetate groups, the polycarbonate

$$\left(O-\!\!\langle\!\!\!\rangle\!\!-\!\!\underset{\underset{CCl_3}{|}}{\overset{\overset{H}{|}}{C}}\!-\!\!\langle\!\!\!\rangle\!\!-\!O-CO\right)_n \qquad (5.4)$$

and polystyrene functionalized by conversion of some units to the trichloroacetate

$$\begin{array}{c} -CH_2-CH- \\ \langle\!\!\!\rangle \\ CH_2-O-\overset{\overset{O}{||}}{C}-CCl_3 \end{array} \qquad (5.5)$$

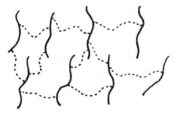

Figure 5.5. An idealized ABCP formed by conterminous grafting. After Bamford and Eastmond.[66]

Virtually the only restriction on the monomer that is used to form the B component of the above type of ABCPs is that it should be polymerizable by free radicals in solution. In most of the synthetic work carried out to date methyl methacrylate, styrene, and chloroprene have been employed, but several other monomers were used in kinetic studies.

The relative quantity of conterminous grafting to simple grafting obtained depends on the ratio of combination to disproportionation, as illustrated in the following pair of reactions:

$$\left\{-(M)_{\ddot{n}} + \cdot(M)_n-\right\} \xrightarrow[\text{disproportionation}]{\text{combination}} \begin{cases} \left\{ \cdots \sim \cdots \right\} \ (I) \\ \\ \left\{ \cdots \cdots \right\} \ (II) \end{cases} \tag{5.6}$$

Under some kinds of reaction conditions, for example, poly(methyl methacrylate) has a combination:disproportionation ratio of about 0.25, while that of polystyrene is close to 10. Thus, polystyrene will yield more of structure I in reaction (5.6), while poly(methyl methacrylate) will yield more of structure II. At light grafting levels, of course, the products will be soluble.

5.6.2. Conjugated Diene ABCPs

Baldwin and Gardner[68] prepared somewhat more complex structures, as illustrated in Figure 5.6. They employed a new elastomer as polymer, designated as a conjugated diene butyl, CDB, and thought to have the following structures:

$$\sim C-\underset{\underset{C}{|}}{\overset{\overset{C}{|}}{C}}-C-\underset{}{\overset{C}{\parallel}}C-C=C-C-\underset{\underset{C}{|}}{\overset{\overset{C}{|}}{C}}\sim \tag{5.7}$$

$$\sim C-\underset{\underset{C}{|}}{\overset{\overset{C}{|}}{C}}-C=\overset{C}{\overset{|}{C}}-C=C-C-\underset{\underset{C}{|}}{\overset{\overset{C}{|}}{C}}\sim \tag{5.8}$$

This elastomer might be considered as a polyfunctional macromonomer and accordingly be employed in radical-initiated copolymerizations with simple mono- and polyfunctional vinyl-type monomers. Since crosslinking of the elastomer by chains derived from the added monomer would be the likely consequence of such a copolymerization, it was clear that the

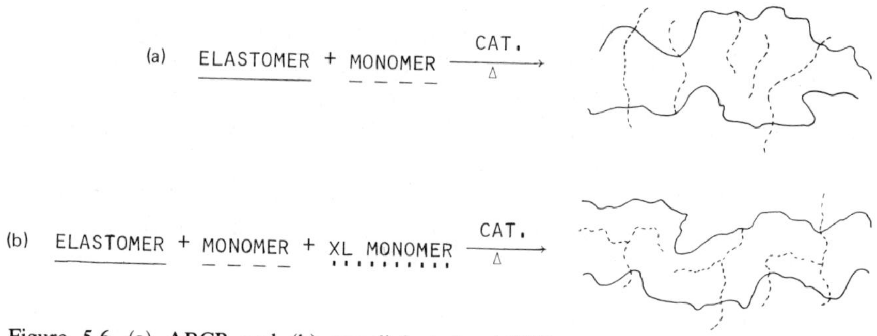

Figure 5.6. (a) ABCP and (b) crosslinked (semi-IPN) structures. After Baldwin and Gardner.[68]

copolymerization would have to take place during a molding operation in order to be useful. The final product structure would be an ABCP.

Two illustrations of this process are shown in Figure 5.6. In the simplest case, i.e., single chains derived from the monomer(s) II grafting onto polymer I in a tetrafunctional linkage, Figure 5.6(a), some conterminous grafting and some free homopolymer would be anticipated, in addition to simple grafts. In a variation employing some polyfunctional monomer along with the monofunctional monomer, the chains derived from the latter would themselves be branched or crosslinked, Figure 5.6(b).

There are two important features of this graft-curing process which make it potentially useful. First, the ultimate level of crosslinking of the polymer I elastomer is controlled by having only limited reactive functionality present. The CDB employed contained some 1.4 mol % diene, about half of which was of structure (5.7). Second, the fact that the "solvent" for monomer II is a viscous elastomer, coupled with the fact that crosslinking occurs during polymerization, automatically reduces the rate of termination of the high-molecular-weight radical chains so that the rate of polymerization is high (Trommsdorff effect).

5.6.3. Castable Polyesters

Two materials of considerable commercial importance make use of the ABCP structure. The first are the castable polyesters,[69-71] usually made from unsaturated polyesters such as polymer I, with styrene serving as monomer II. The unsaturated polyesters are based on glycols and dibasic acids. The dibasic acid portion is composed of one or more saturated acids and an unsaturated acid.[72] The unsaturated acid is usually maleic or fumaric acid, while the saturated acid may be phthalic, adipic, etc. The polymer is dissolved in styrene monomer. Subsequent peroxide-initiated free-radical polymerization forms a three-dimensional ABCP network.

5.6.4. Epoxy/CTBN Materials

The second ABCP-type materials of commerce are the rubber-toughened epoxy resins.[73-78] Usually the epoxy resins are based on the diglycidyl ether of bisphenol A (DGEBA). Carboxyl-terminated butadiene–acrylonitrile (CTBN) rubber is the elastomer of preference. A typical composition is[77]

$$HOOC \text{-}[(CH_2\text{-}CH\text{=}CH\text{-}CH_2)_x \text{-}(CH_2\text{-}\underset{\underset{CN}{|}}{CH})_y]_m COOH \qquad (5.9)$$

where $x = 5$, $y = 1$, and $m = 10$, giving an average molecular weight of 3320. A typical recipe is 100 DGEBA epoxy resin, 10 CTBN rubber, 6 piperidine.

The reaction first involves salt formation between the carboxyl and the amine:

$$\underset{\substack{\|\\O}}{R'\text{-}C}\text{-}OH + R_3N \rightarrow \underset{\substack{\|\\O}}{R'\text{-}C}\text{-}\overset{\ominus}{O} \quad \underset{CH_2\text{-}CH_2}{\overset{H}{\underset{}{\overset{}{\text{-}N}}}} \overset{CH_2\text{-}CH_2}{\underset{CH_2\text{-}CH_2}{\diagup\diagdown}} CH_2 \qquad (5.10)$$

The carboxylate salt then reacts quite rapidly with the epoxy group:

$$\underset{\substack{\|\\O}}{R'\text{-}C}\text{-}\overset{\ominus}{O}\text{-}R_3\overset{\oplus}{N}H + \overset{O}{\overset{\diagdown}{CH_2\text{-}CH}}\text{-}E \rightarrow R'\text{-}\underset{\substack{\|\\O}}{C}\text{-}O\text{-}CH_2\text{-}\underset{|}{CH}\text{-}E \qquad (5.11)$$

Since the CTBN and DGEBA are reacted end-on-end-on-end, an ABA block copolymer is formed. Of course, the ordinary epoxide cross-linking reactions are taking place at the same time, so that, in effect, the linear CTBN is joined conterminously with the epoxy network, forming an ABCP–semi-IPN.

5.7. THERMOPLASTIC IPNs

While all of the materials discussed above contain standard covalent crosslinks, it is also possible to utilize a range of physical crosslinking modes.* Three distinct types of physical crosslinks may be recognized.

---

* No special topological designation can be invoked.

a. *Block Copolymers.* Materials of the ABA type, known as thermoplastic elastomers,* fall into this category. Frequently the end (A) blocks are made of polystyrene, while the central (B) block is composed of polybutadiene, polyisoprene, or ethylene–butylene rubber. The end blocks form a discrete phase, physically crosslinking the system.[79,80]

b. *Ionomer Formation.* Ionic groups such as sodium methacrylate, may be introduced along the polymer chain. At a level of about 5 mol %, a segregation takes place, and the ionic groups form small, concentrated domains.[81]

c. *Partially Crystalline Polymers.* It has long been recognized that crystalline regions (surrounded by amorphous material) serve as physical crosslinks. A well-known example is polyethylene, where the crystalline regions cause the whitish haze. The flexible chains surrounding the crystallites impart extensive deformability to the material.[82]

Interesting enough, each of the above materials behaves as a thermoplastic at elevated temperatures, i.e., they are capable of being processed. Combinations of these materials form the basis for the newest category of IPNs: thermoplastic IPNs. Obviously, they offer features heretofore unavailable in IPNs. While the SINs could be processed up to the gelation stage, the thermoplastic IPNs can be processed or reprocessed at any time.

Each of the above physical crosslinking modes can be employed in IPN formation, using either the same or a different physical crosslinking mode. In addition, frequently a deliberate effort is made to achieve dual-phase continuity. Two major routes have been employed to synthesize these materials: (1) mechanical blending of the two polymers in the "melt" stage,[83,84] and (2) chemical blending, by swelling monomer II into polymer I (or dissolving polymer I in monomer II), and polymerizing II *in situ.*[85]

For the mechanical-blending method, six combinations of blocks, ionomers, and crystalline polymers are possible. The chemical-blending route, which distinguishes the order of polymerization, yields nine possible combinations.

An example of the synthesis of a chemically blended thermoplastic IPN[85] was the preparation of a poly(styrene-*b*-ethylene-co-butylene-*b*-styrene) (SEBS)/poly(styrene-*co*-sodium methacrylate) composition. Styrene and methacrylic acid (90/10) was swollen into the triblock copolymer and thermally polymerized with benzoyl peroxide. The mass was placed in a Brabender and heated. A stoichiometric quantity of sodium hydroxide was slowly added in aqueous solution. The water flashes off as steam, leaving the thermoplastic IPN.

---

* Segmented polyurethanes (TPU) and thermoplastic polyester elastomers such as "Hytrel" by du Pont are also important thermoplastic elastomers.

Numerous crystalline/crystalline mechanical blends have been prepared. Very recently, for example, Cruz et al.[86] studied a series of polyester/polycarbonate blends, distinguished by two aspects: (1) Both polymers exhibited semicrystallinity, thus forming a thermoplastic IPN, and (2) the amorphous portions of both phases were mutually soluble.

Plochocki[87] has reviewed the melt blending and rheology of polyolefin blends, citing 52 patents. The more practical aspects of the thermoplastic IPNs are discussed in Section 8.7.

REFERENCES

1. L. H. Sperling and D. W. Friedman, *J. Polym. Sci. A-2* **7**, 425 (1969).
2. L. H. Sperling, H. F. George, V. Huelck, and D. A. Thomas, *J. Appl. Polym. Sci.* **14**, 2815 (1970).
3. A. J. Curtius, M. J. Covitch, D. A. Thomas, and L. H. Sperling, *Polym. Eng. Sci.* **12**, 101 (1972).
4. L. H. Sperling, V. Huelck, and D. A. Thomas, in *Polymer Networks: Structural and Mechanical Properties*, A. J. Chompff and S. Newman, eds., Plenum, New York (1971).
5. V. Huelck, D. A. Thomas, and L. H. Sperling, *Macromolecules* **5**, 340, 348 (1972).
6. A. A. Donatelli, L. H. Sperling, and D. A. Thomas, in *Recent Advances in Polymer Blends, Grafts, and Blocks*, L. H. Sperling, ed., Plenum, New York (1974).
7. A. A. Donatelli, L. H. Sperling, and D. A. Thomas, *Macromolecules* **9**, 671, 676 (1976).
8. G. M. Yenwo, J. A. Manson, J. Pulido, L. H. Sperling, A. Conde, and N. Devia, *J. Appl. Polym. Sci.* **21**, 153 (1977).
9. G. M. Yenwo, L. H. Sperling, J. Pulido, J. A. Manson, and A. Conde, *Polym. Eng. Sci.* **17**, 251 (1977).
10. G. M. Yenwo, L. H. Sperling, J. A. Manson, and A. Conde, in *Chemistry and Properties of Crosslinked Polymers*, S. S. Labana, ed., Academic, New York (1977).
11. Yu. S. Lipatov, L. M. Sergeeva, L. M. Mozzukhina, and N. P. Apukhtina, *Vysokomol. Soedin.* **A16**(10), 2290 (1974).
12. Yu. S. Lipatov, L. M. Sergeeva, L. M. Karabanova, A. Ye Nesterov, and T. D. Ignatova, *Vysokomol. Soedin.* **A18**(5), 1025 (1976).
13. T. E. Lipatova, V. V. Shilov, N. P. Bazilevskaya, and Yu. S. Lipatov, *Br. Polym. J.* **9**, 159 (1977).
14. H. L. Frisch, D. Klempner, and K. C. Frisch, *Polym. Lett.* **7**, 775 (1969).
15. D. Klempner, H. L. Frisch, and K. C. Frisch, *J. Polym. Sci. A-2* **8**, 921 (1970).
16. M. Matsuo, T. K. Kwei, D. Klempner, and H. L. Frisch, *Polym. Eng. Sci.* **10**, 327 (1970).
17. D. Klempner and H. L. Frisch, *Polym. Lett.* **8**, 525 (1970).
18. D. Klempner, H. L. Frisch, and K. C. Frisch, *J. Elastoplast.* **3**(1), 2 (1971).
19. L. H. Sperling, T. W. Chiu, and D. A. Thomas, *J. Appl. Polym. Sci.* **17**, 2443 (1973).
20. L. H. Sperling, T. W. Chiu, C. Hartman, and D. A. Thomas, *Int. J. Polym. Mater.* **1**, 331 (1972).
21. L. H. Sperling, T. W. Chiu, R. G. Gramlich, and D. A. Thomas, *J. Paint Technol.* **46**, 47 (1974).
22. L. H. Sperling, in *Noise-Con 73 Proceedings*, D. R. Tree, ed., Institute of Noise Control Engineering, Poughkeepsie, New York (1973).

23. J. A. Grates, D. A. Thomas, E. C. Hickey, and L. H. Sperling, *J. Appl. Polym. Sci.* **19**, 1731 (1975).

24. L. H. Sperling, D. A. Thomas, J. E. Lorenz, and E. J. Nagel, *J. Appl. Polym. Sci.* **19**, 2225 (1975).

25. J. E. Lorenz, D. A. Thomas, and L. H. Sperling, in *Emulsions Polymerization*, I. Purma and J. L. Gardon, eds., American Chemical Society, Washington, D.C. (1976).

26. L. H. Sperling and D. A. Thomas, U.S. Pat. 3,833,404 (1974).

27. J. Sionakidis, L. H. Sperling, and D. A. Thomas, *J. Appl. Polym. Sci.* **24**, 1179 (1979).

28. M. Matsuo, *Jpn. Plast.* **2**, (July) 6, 1968.

29. J. Khandheria and L. H. Sperling, unpublished.

30. C. Placek, Chemical Process Review No. 46, Noyes Data Corp. Park Ridge, New Jersey 1970.

31. C. B. Bucknall, *Toughened Plastics*, Applied Science, London (1977).

32. R. A. Dickie, M. F. Cheung, and S. Newman, *J. Appl. Polym. Sci.* **17**, 65 (1973).

33. L. H. Sperling and R. R. Arnts, *J. Appl. Polym. Sci.* **15**, 2731 (1971).

34. R. E. Touhsaent, D. A. Thomas, and L. H. Sperling, *J. Polym. Sci.* **46C**, 175 (1974).

35. R. E. Touhsaent, D. A. Thomas, and L. H. Sperling, in *Toughness and Brittleness of Plastics*, R. D. Deanin and A. M. Crugnola, eds., Advances in Chemistry Series No. 154, American Chemical Society, Washington, D.C. (1976).

36. S. C. Kim, D. Klempner, K. C. Frisch, H. L. Frisch, and H. Ghiradella, *Polym. Eng. Sci.* **15**, 339 (1975).

37. S. C. Kim, D. Klempner, K. C. Frisch, W. Radigan, and H. L. Frisch, *Macromolecules* **9**, 258 (1976).

38. S. C. Kim, D. Klempner, K. C. Frisch, and H. L. Frisch, *Macromolecules* **9**, 263 (1976).

39. S. C. Kim, D. Klempner, K. C. Frisch, and H. L. Frisch, *Macromolecules* **10**, 1187 (1977).

40. S. C. Kim, D. Klempner, K. C. Frisch, and H. L. Frisch, *Macromolecules* **10**, 1191 (1977).

41. S. C. Kim, D. Klempner, K. C. Frisch, and H. L. Frisch, *J. Appl. Polym. Sci.* **21**, 1289 (1977).

42. N. Devia-Manjarres, J. A. Manson, L. H. Sperling, and A. Conde, *Polym. Eng. Sci.* **18**, 200 (1978).

43. N. Devia, J. A. Manson, L. H. Sperling, and A. Conde, *Macromolecules* **12**, 360 (1979).

44. N. Devia, J. A. Manson, L. H. Sperling, and A. Conde, *Polym. Eng. Sci.* **19**(12), 869 (1979).

45. N. Devia, J. A. Manson, L. H. Sperling, and A. Conde, *Polym. Eng. Sci.* **19**(12), 878 (1979).

46. N. Devia, J. A. Manson, L. H. Sperling, and A. Conde, *J. Appl. Polym. Sci.* **24**, 569 (1979).

47. K. C. Frisch, D. Klempner, T. Antzak, and H. L. Frisch, *J. Appl. Polym. Sci.* **18**, 683 (1974).

48. H. L. Frisch, K. C. Frisch, and D. Klempner, *Polym. Eng. Sci.* **14**, 646 (1974).

49. K. C. Frisch, D. Klempner, S. Migdal, H. L. Frisch, and H. Ghiradella, *Polym. Eng. Sci.* **14**, 76 (1974).

50. K. C. Frisch, D. Klempner, S. Migdal, and H. L. Frisch, *J. Polym. Sci. Polym. Chem. Ed.* **12**, 885 (1974).

51. K. C. Frisch, D. Klempner, S. Migdal, H. L. Frisch, and A. P. Dunlop, *J. Appl. Polym. Sci.* **19**, 1893 (1975).

52. K. C. Frisch, D. Klempner, H. L. Frisch, and H. Ghiradella, in *Recent Advances in Polymer Blends, Grafts, and Blocks*, L. H. Sperling, ed., Plenum, New York (1974).

53. K. C. Frisch, D. Klempner, S. K. Mukherjee, and H. L. Frisch, *J. Appl. Polym. Sci.* **18**, 689 (1974).

54. P. R. Scarito and L. H. Sperling, *Polym. Eng. Sci.* **19**, 297 (1979).

55. D. Klempner, H. K. Yoon, K. C. Frisch, and H. L. Frisch, in *Chemistry and Properties of Crosslinked Polymers*, S. S. Labana, ed., Academic, New York (1977).
56. B. V. Ozerkovskii, Yu B. Kalmykov, U. G. Gafunov, and V. P. Roshchupkin, *Vysokomol. Soedin.* **A19**(7), 1437 (1977).
57. G. Allen, M. J. Bowden, D. J. Blundell, F. G. Hutchinson, G. M. Jeffs, and J. Vyvoda, *Polymer* **14**, 597 (1973).
58. G. P. Belonovskaya, J. D. Chernova, L. A. Korotneva, L. S. Andrianova, B. A. Dolgoplosk, S. K. Zakharov, Yu S. Sazanov, K. K. Kalninsh, L. M. Kaljuzhnaya, and M. F. Lebedeva, *Eur. Polym. J.* **12**, 817 (1976).
59. L. H. Sperling, *Polym. Prepr. Am. Chem. Soc. Div. Polym. Chem.* **14**, 958 (1973).
60. R. E. Touhsaent, D. A. Thomas, and L. H. Sperling, *J. Polym. Sci. Symp.* **46**, 175 (1974).
61. A. F. Wells, *Structural Inorganic Chemistry*, 4th ed., Clarendon, Oxford (1975).
62. G. C. Meyer and P. Y. Mehrenberger, *Eur. Polym. J.* **13**, 383 (1977).
63. G. C. Eastmond and E. G. Smith, *Polymer* **17**, 367 (1976).
64. C. H. Bamford, G. C. Eastmond, and D. Whittle, *Polymer* **16**, 377 (1975).
65. G. C. Eastmond and D. G. Phillips, in *Polymer Alloys*, D. Klempner and K. C. Frisch, eds., Plenum, New York (1977).
66. C. H. Bamford and G. C. Eastmond, in *Recent Advances in Polymer Blends, Grafts, and Blocks*, L. H. Sperling, ed., Plenum, New York (1974).
67. C. H. Bamford, G. C. Eastmond, and D. Whittle, *Polymer* **12**, 247 (1971).
68. F. P. Baldwin and I. J. Gardner, in *Chemistry and Properties of Crosslinked Polymers*, S. S. Labana, ed., Academic, New York (1977).
69. E. J. Bartkus and C. H. Kroekel, in *Polyblends and Composites*, P. F. Bruins, ed., Wiley, New York (1970).
70. F. Fekete and J. S. McNally, Fr. Pat. 1,567,700 (1969).
71. D. Katz and A. V. Tobolsky, *J. Polym. Sci. A-2* **2**, 1587 (1964).
72. H. P. Cordts and J. A. Bauer, in *Modern Plastics Encyclopedia*, McGraw-Hill, New York (1978).
73. C. B. Bucknall and T. Yoshi, *Br. Polym. J.* **10**, 53 (1978).
74. A. C. Soldatos and A. S. Burhans, in *Advances in Chemistry Series No. 99*, N.A.J. Platzer, ed., American Chemical Society, Washington, D.C. (1971).
75. N. K. Kalfoglou and H. L. Williams, *J. Appl. Polym. Sci.* **17**, 1377 (1973).
76. J. N. Sultan and F. J. McGarry, *Polym. Eng. Sci.* **13**, 29 (1973).
77. C. B. Bucknall, *Toughened Plastics*, Applied Science, London (1977), pp. 33–35.
78. E. H. Rowe, A. R. Siebert, and R. S. Drake, *Mod. Plast.* **49**(9), 110 (1970).
79. S. L. Aggarwal, *Polymer* **17**, 838 (1976).
80. A. Noshay and J. E. McGrath, *Block Copolymers, Overview and Critical Survey*, Academic, New York (1977).
81. R. H. Kinsey, *Appl. Polym. Symp.* **11**, 77 (1969).
82. P. H. Geil, *Polymer Single Crystals*, Interscience, New York (1963).
83. E. N. Kresge, in *Polymer Blends*, D. R. Paul and S. Newman, eds., Academic, New York (1978), Chap. 20.
84. A. L. Bull and G. Holden, *J. Elast. Plast.* **9**(7), 281 (1977).
85. D. L. Siegfried, D. A. Thomas, and L. H. Sperling, *Polymer Preprints* **21(1)**, 186 (1980).
86. C. A. Cruz, D. R. Paul, and J. W. Barlow, *J. Appl. Polym. Sci.* **23**, 589 (1979).
87. A. P. Plochocki, in *Polymer Blends*, Vol. 2, D. R. Paul and S. Newman, eds., Academic, New York (1978).

6

MORPHOLOGY AND GLASS TRANSITION BEHAVIOR

6.1. INTRODUCTION

Because of their dual crosslinked nature, both networks exert a unique control over the size, shape, and composition of the phase domains in an IPN. The morphological detail strongly influences, in turn, the physical and mechanical behavior of the material. While Chapter 5 detailed several ways of synthesizing IPNs, little mention was made of how crosslink density, order of polymerization, overall composition, etc. affect the final product. The objective of this chapter will be to explore the interrelationships among synthesis, morphology, and glass transition behavior. Mechanical and engineering properties will be treated in Chapter 7.

In an important way, this chapter also picks up where Chapter 2 left off. Chapter 2 described the morphology and mechanical behavior of several classes of polymer blends, grafts, and blocks. Like these materials, covalent bonding between the species encourages molecular mixing. Crosslinking plays a similar role. The nature of the interphase material requires special attention. It must be emphasized at the outset that all of these materials are interesting and useful because of their complex structure, certainly not in spite of this fact.

The behavior of multicomponent polymer systems has been the subject of several recent reviews. Bucknall[1] has examined polymer blends and grafts, with particular reference to the toughening of plastics. All aspects of block copolymers have been explored by Noshay and McGrath.[2] Manson and Sperling[3] reviewed the entire field of polymer blends and composites.

Recent reviews in the field of IPNs have tended to emphasize the relationship between morphology and behavior, the subject of the present chapter. These reviews include materials by Lipatov and Sergeeva,[4] Klempner,[5] Sperling,[6] and Frisch et al.[7] The edited two-volume work by Paul and Newman[8] provides the most recent coverage of the entire field of multipolymer materials and includes a review of IPN materials.[9]

6.2. MOLECULAR CONTROL OF MORPHOLOGY

Most IPNs and related materials investigated to date show phase separation. The phases, however, vary in amount, size, shape, sharpness of their interfaces, and degree of continuity. These aspects together constitute the morphology of the material, and the multitude of possible variations in the morphology controls many of the material properties.

Some aspects of morphology can be observed directly by transmission electron microscopy of stained and ultramicrotomed thin sections.[10-12] The most successful staining method, developed by Kato, makes use of osmium tetroxide, which attacks the double bonds in diene type polymers.[10] The OsO$_4$ staining technique can also be used with other active groups, such as polyurethanes.[5] Many saturated or nonreactive polymers are not easily studied by transmission electron microscopy, unfortunately, because they cannot be stained. Other aspects of morphology, such as phase continuity and interface characteristics, are best determined by combining chemical and dynamic mechanical spectroscopy methods with electron microscopy.

Some of the factors that control the morphology of IPNs are now reasonably clear; they include chemical compatibility of the polymers, interfacial tension, crosslink densities of the networks, polymerization method, and the IPN composition. While these factors may be interrelated, they can often be varied independently. Their effects are summarized here.

A degree of compatibility between polymers may be brought about by IPN formation, because the two polymers are interlocked in a three-dimensional structure imposed by the synthetic method. Phase separation

Table 6.1. Solubility Parameters of Selected Monomers and Polymers

Structure	δ, Solubility parameter $(cal/cm^3)^{1/2}$	
	Monomer	Polymer
Poly(vinyl chloride)	7.8	10
1,3-Polybutadiene	7.1	8
Polyacrylonitrile	10.5	13
Polystyrene	9.3	9.1
Poly(ethyl acrylate)	8.6	9.4
Poly(methyl methacrylate)	8.8	9.5
Castor oil polyester	—	9.3
Polyurethane	—	10
Castor oil polyurethane	—	9.2
Polyisoprene	7.4	8.1

generally ensues in the course of polymerization, and the resulting phase domain size is smaller for higher compatibility systems[13] than for lower compatibility systems, and usually much smaller than obtainable by mechanical blending techniques; see Chapter 2.

Although the term "compatibility" has been used to indicate wettability or the compatibility of forming a monolithic solid, the term also has an important thermodynamic connotation. Assuming equilibrium conditions, a compatible polymer mixture ought to form a true solution. The free energy of mixing, then, must be zero or negative. Some aspects of this were considered in Chapter 2. A practical guide to polymer–polymer compatibility is the solubility parameter, δ; see Table 6.1.[14] While the heat of mixing is influenced by polarity considerations, a similar value of δ usually indicates better compatibility than if the values are further apart. In the following, the term "semicompatibility" will refer to materials which mix on a molecular or near molecular scale (50–300 Å), and the term "incompatible" will be reserved for mixtures which exhibit coarser morphologies.*

Other practical indications of compatibility include clarity and a single glass transition temperature. Clarity, or optical transparency, usually indicates very small or nonexistent phase domains, although this can be misleading if the refractive indices of the two polymers match. A single glass transition yields good evidence of molecular mixing, provided that the glass transitions of the homopolymers are different and the glass transition of the mixture is intermediate between the two.

6.3. MORPHOLOGY VIA OPTICAL MICROSCOPY

Microscopy can be done at two levels. Optical microscopy, with a magnification range of up to about 1000, reveals coarse structures and is useful as a screening tool to separate the several types of morphology. Transmission electron microscopy, with magnifications of 100,000 times, can be used to study the several morphological details.

Usually, it is a good idea to study the morphology by optical microscopy first, in any case. An excellent series of optical photomicrographs was presented by Ozerkovskii et al.[15] Figure 6.1[15] shows a series of butyl rubber–polyacrylate blends, semi-SINs, and an SIN. The synthesis of these materials was already summarized in Table 5.7. Figures 6.1a, 6.1d, and 6.1g show the behavior of BR/PMA blends. Figure 6.1a shows a matrix of BR with included spherical particles of PMA. The material passes through a

---

* O. Olabisi, L. M. Robeson, and M. T. Shaw, in their book "Polymer–Polymer Miscibility," Academic Press, New York, 1979, prefer the term "miscibility" to "compatibility," especially when used in the scientific sense of indicating mixing on a molecular level.

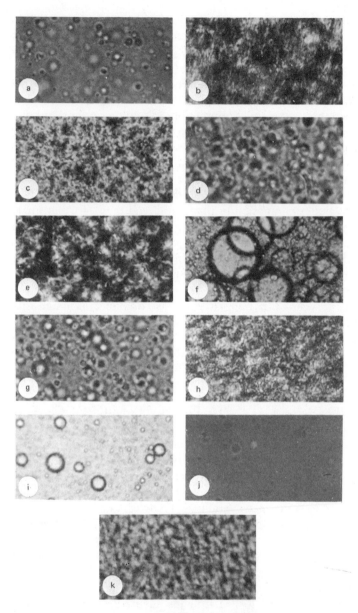

Figure 6.1. (a)–(i): The photomicrographs of BR (butyl rubber)–polyacrylate blends containing (%w/w): a–c, 10; d–f, 50; g–i, 90 of PMA, PNA, and PBA, respectively, (×650). (j), (k): The photomicrographs of PNA–BR (50%) SINs produced at 70°C by polymerizing nonyl acrylate in the presence of 2% benzoyl peroxide and simultaneous vulcanization of BR with 0.3% dinitrobenzene (3a), and without vulcanization (3b) (×650).[15]

phase inversion between Figure 6.1d and 6.1g. Figures 6.1b, 6.1e, and 6.1h present an entirely different situation. Neither of the components forms isolated particles and all three compositions are characterized by a considerable interfacial surface area. This arises from the relatively low interfacial tensions, see Table 5.7.

Figure 6.1j illustrates the SIN. It shows a fairly homogeneous polymeric body, in which relatively small, isolated particles are included. However, when the BR is not vulcanized, Figure 6.1k, phase separation is more pronounced, and the composite forms a well-defined, heterogeneous system.

6.4. MORPHOLOGY VIA TRANSMISSION ELECTRON MICROSCOPY

When the structure cannot be characterized by optical microscopy, transmission electron microscopy comes to the fore, although scanning electron microscopy and or replica techniques are also often helpful.[16] Most of the work to date utilizes osmium tetroxide as a stain. Samples that cannot be stained by osmium tetroxide or another similar agent often cannot be studied by electron microscopy, because the phases cannot be distinguished. Sometimes, a double bond is deliberately added in small quantities during the synthesis to facilitate staining with osmium tetroxide. This is easy to do in acrylic or styrene based systems with an addition of a trace of butadiene or isoprene.

6.4.1. Sequential IPNs

Using the synthesis described in Section 5.2.1, Huelck et al.[17] varied compatibility systematically in sequential IPNs, with poly(ethyl acrylate) (PEA) as polymer I and copolymers of methyl methacrylate (MMA) and styrene (S) as polymer II; see Figure 6.2. For PEA/PMMA, in which the components are isomeric and nearly compatible, dispersed phase domains less than 100 Å in size (fine structure) are found (Figure 6.2a). Such a fine structure occurs because the high compatibility (ΔG_M just above zero) precludes phase separation until a high conversion to PMMA. Referring to Table 6.1, it is observed that the solubility parameters of PEA and PMMA are significantly closer together than the values of PEA and PS. As developed in Chapter 5, in fact, PEA and PMMA are chemically isomeric.

With the much less compatible system PEA/PS (Figure 6.2b), a cellular structure of about 1000 Å in size is found. Here, phase separation is thought to take place earlier in the polymerization, forming cellular regions rich in

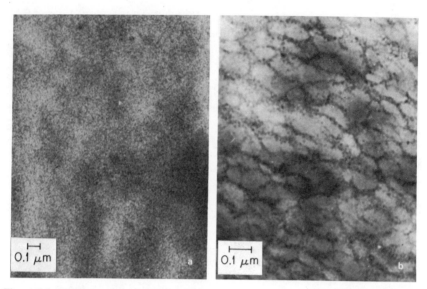

Figure 6.2. Electron micrographs of IPN's of (a) 75/25 poly(ethyl acrylate)/poly(methyl methacrylate), and (b) 50/50 poly(ethyl acrylate)/polystyrene. A small amount of butadiene was copolymerized with the poly(ethyl acrylate) to aid in osmium tetroxide staining.

PS and S monomer. A second phase separation follows, leading to a fine structure of PS domains in the PEA cell walls as well as the cellular structure itself.

Styrene–butadiene rubber/polystyrene (SBR/PS) IPNs are relatively incompatible, even though they are both nonpolar polymers and their solubility parameters differ by only about 1.0; see Table 6.1. They show distinct phase separation and a cellular domain structure as discussed earlier; see Figure 2.3, lower left. The most interesting feature of the morphology illustrated in Figure 2.3 is the apparent dual phase continuity.[18] When polymer II is present in the greater weight concentration, it is speculated that the cell wall structures present in Figure 6.2b open, and this open-celled structure sometimes exhibits the apparent dual-phase continuity shown in Figure 2.3.

The cellular structures illustrated in Figures 6.2 and 2.3 (lower two photos) call forth several further comments. First, the organization of the cellular structures, especially those shown in Figure 2.3, is reminiscent of the morphology seen within the rubber droplets in graft-type high-impact polystyrene; see Figure 2.3, upper left. However, in the case of the IPNs, the cellular structures pervade the entire sample. In part, the similarity may stem from the fact that the rubber portion of HiPS is itself being crosslinked during or prior to polymerization II.[1] (Actually, an AB crosslinked polymer, $P_1 O_C P_2$, is being formed.) Indeed, cellular structures of 500–2000 Å may be peculiar to the formation of polymer II domains within a crosslinked polymer I.

Second, in the sequential IPNs, the volume occupied by polymer I is increased by the swelling action of monomer II, so that the Flory–Rehner elastic forces are called into play; see Chapter 4. In the final product, polymer I chains are always elastically extended out of their relaxed conformation. Besides limiting the total amount of monomer II that can be imbibed, the crosslinking level in polymer I participates in controlling the domain size of polymer II, as will be described below and in Section 6.5.

An additional comment must be made about the fine structure observed in Figure 6.2a. Indeed, Donatelli et al.[18] also observed this structure in high-magnification electron micrographs of the materials shown in Figure 2.3. This phenomenon appears to be general and is most likely caused by a late phase separation in the cell walls comprising the polymer-I-rich phase during polymerization. This can be seen in Figure 6.2b.

By way of summary, the morphology of incompatible sequential IPNs develops as follows:

1. Monomer II plus crosslinker and activator is swollen into network I to produce a uniform, one-phase composition. The free-energy change on mixing is negative.

2. As polymerization of monomer II proceeds, the free energy of mixing goes from negative to positive. Phase separation ensues, developing a cellular structure with polymer I as the cell walls, and polymer II as the cell contents. At this stage both polymers are highly swollen with monomer II. The cellular structures are of the order of 1000 Å in diameter.

3. As polymerization of monomer II continues, a second phase separation occurs, resulting in the 100-Å fine structure.

It is not completely understood however, whether the fine structure associated with semicompatibility and that arising from the latter portions of an incompatible IPN polymerization arise from the same or different thermodynamic origins.

6.4.1.1. Effect of Crosslinking Level

In the case of block and graft copolymers, the compatibility is known to be improved by the presence of intermolecular bonds. For block copolymers, quantitative relationships have already been established, putting the free energy of mixing on one side of an equation, and the concentration of A–B bonds on the other.[19,20] This subject has recently been reviewed.[2,3]

Because of the qualitative similarity between a crosslink and a graft or block site, the crosslinks in an IPN are thought to play a similar thermodynamic role. Thus, more crosslinks in an IPN must mean a smaller positive (or more negative) value of the free energy of mixing. Although the relationships have only been partially worked out, this chapter will provide some of the growing evidence of the role that crosslink concentration plays in the thermodynamics of mixing.

For example, increased crosslink density in polymer network I in an IPN clearly decreases the domain size of polymer II.[18] This is illustrated by comparison of Figure 2.3, bottom left and bottom right. This effect appears reasonable because a tighter initial network must restrict the size of the regions in which polymer II can phase separate. However, the role of crosslinks in merely diminishing phase domain size, and in increasing compatibility in a thermodynamic sense, needs to be carefully distinguished.

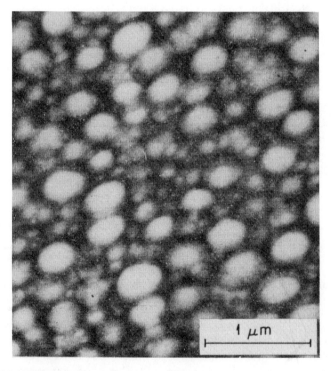

Figure 6.3. Electron micrograph of an "interstitial polymer" of 80/20 polyurethane/poly(methyl methacrylate). Poly (butadiene diols) in the polyurethane permit staining with osmium tetroxide.[23]

The effect has also been rationalized by a semiempirical thermodynamic model[21]; see Section 6.5.2.

Variation of crosslink density in the second network has little effect on the IPN morphology, surprisingly, indicating that the first network is controlling.[21] In the extreme case of a linear polymer II, the morphologies were somewhat less uniform.[18,21] With no crosslinking in polymer I, on the other hand, it remained continuous but both phases were much coarser and irregular. Further discussion on this material is associated with Figure 2.3.

6.4.1.2. Related Materials

Allen et al.[22,23] observed similar morphological effects in polyurethane PU/PMMA materials, made by "interstitial polymerization" (see Figure 6.3). The morphology shows roughly spherical domains of PMMA in a matrix of PU.[22] The PU network formed first from a solution of reactants including the MMA, but reaction conditions permitted varying tightness of the network; see Section 5.5.2.4. Although the PMMA was linear, its domain size varied from about 650 Å for tight PU networks to about 1800 Å for loose networks.[23]

The AB crosslinked copolymers of Bamford et al.[24] are not strictly IPNs, since the polymer II chains are part of the same network as the polymer I chains. However, like graft copolymers and IPNs, they do phase separate into distinct domains.[25,26] The polymer II chains usually form the discontinuous phase, because, similarly to the sequential IPNs, they are not stirred during polymerization.

6.4.2. Simultaneous Interpenetrating Networks

6.4.2.1. Polyurethane SINs

For sequential IPNs the effects of compatibility and crosslinking have already been described. When the order of polymerization is reversed in sequence, however, the new morphology is again controlled principally by the first network.[17] In the synthesis of SINs, both networks form at the same time, but not necessarily at the same rate (see Section 5.5).

The effect of morphology as a function of mutual solubility was explored by Frisch et al.[13,27] for polyurethane-based SINs. Two such morphologies are shown in Figure 6.4.[13,27] Figure 6.4a illustrates the morphology of a polyurethane/poly(methyl methacrylate) pair, which has a closer match in solubility parameters than the polyurethane/polystyrene SIN shown in Figure 6.4b; see Table 6.1. The PU/PMMA composition exhibits significantly greater molecular mixing than the PU/PS composition.

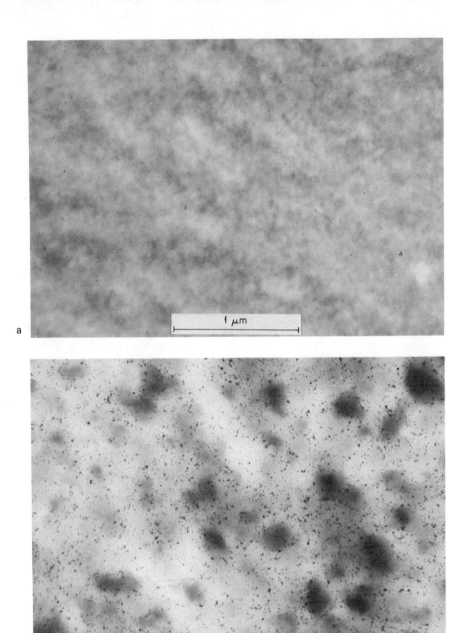

Figure 6.4. Morphology of simultaneous interpenetrating networks. (a) Electron micrograph of SIN of 75/25 polyurethane/poly (methyl methacrylate). Polyurethane stained with osmium tetroxide.[27] (b) 75/25 PU/PS.[13]

6.4.2.2. Castor Oil SINs

Devia et al.[29] explored the effects of stirring on phase inversion in a series of castor oil polyester/polystyrene SINs (see Section 5.5.2.2). For materials subjected to a somewhat different polymerization sequence, Devia et al.[28–32] showed that quite dissimilar morphologies may ensue. The phase inversion of the castor oil–sebacic acid/polystyrene SIN[28–32] provides insight into the labyrinth of polymerization sequence.

Referring back to Figure 5.3, the polyester prepolymer and the styrene–divinyl benzene solution are mixed and heated to 80°C.[29] Samples prepared with rapid stirring and poured into test tubes at different times showed the sequence illustrated schematically in Figure 6.5. The two layers were distinguishable because of dullness and hardness differences. The upper layer possessed more elastomer, which formed the continuous phase, and the bottom layer had the PS as the continuous phase. The volume of the upper layer (elastomer continuous) decreased slowly and finally disappeared after about 90 min of continuous stirring. Samples of both top and bottom layers were studied by transmission electron microscopy techniques, and micrographs for a 10/90 COPE/PSN are shown in Figure 6.6.[29] During this time, a phase inversion took place. Micrographs T2A, T2B, and T2C in Figure 6.6 were all taken from the top layer and illustrate the details of the phase inversion. At T2A the castor oil elastomer (stained dark by the OsO_4) formed the continuous phase. Micrograph T2B shows the polystyrene domain coalescence process, by which elastomer domains were coalesced and a continuous polystyrene phase formed. Thus, micrograph T2B illustrates the actual phase inversion point. Micrographs T2C and B2 show that at the end of the phase inversion the top and bottom regions had identical morphologies and the sample attained macroscopic homogeneity.

In the time interval between phase inversion and gelation of the polystyrene, the final morphological features, such as average size and size distribution of elastomer domains, became fixed. Since these morphological changes affect properties such as modulus and impact resistance, the characteristics of the system just after phase inversion but before gelation demand the closest scrutiny. The open time interval was found to decrease as the polyester prepolymer content was increased, probably because higher

Figure 6.5. Layering effect during the synthesis of a 10/90 COPEN/PSN SIN.[29] The times given are the reaction times with stirring prior to pouring. The shading of the columns is proportional to the softness.

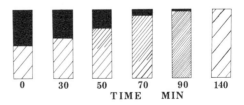

0   30   50   70   90   140

TIME   MIN

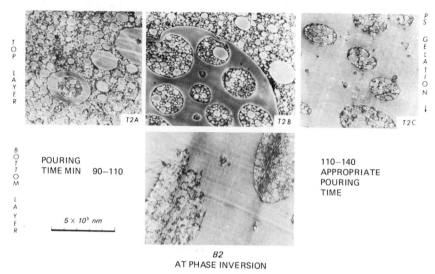

Figure 6.6. Morphology changes induced by stirring during the synthesis of a 10/90 COPE/PSN SIN. Sample poured into the mold at the phase inversion point.[29]

polystyrene conversions were required for the system to reach suitable phase inversion conditions.

The phase-within-a-phase-within-a-phase morphology, depicted in the right-hand portion of Figure 6.6 after phase inversion, looks much like that of high-impact polystyrene prepared by the graft technique (see Figure 2.3). In both, polystyrene forms the matrix, as well as the contents of the cellular structure within the rubber (castor oil) droplets.

Bucknall showed that phase domain sizes of the order of 0.5 μ yield a maximum in impact strength in rubber toughened plastics.[1,33] It is interesting to speculate whether the phase domain size and organization criteria required for toughening will be as important for SIN as for graft copolymers.

Although the morphology found by Kim et al., Figure 6.4, is intuitively expected for a SIN, even having simultaneous gelation of both polymers does not necessarily guarantee molecular mixing or dual-phase continuity.[34,35]

6.4.3. Latex IENs and IPNs

Emulsion polymerization offers two distinct routes to IPNs, depending on whether the second polymer is synthesized as a separate latex, leading to a specialized mixed latex material, or the second polymer is synthesized on the first, leading to core–shell formation. Each exhibits a very distinctive morphology.

6.4.3.1. IENs

Matsuo *et al.*[36] studied the morphological features of poly-acrylate/polyurethane IENs prepared by the mutual coagulation and subsequent crosslinking of the two latexes, as described in Section 5.3. Figure 6.7[36] shows rather coarse phase domains, with the latex particles (stained dark) still identifiable. Although it would appear that the acrylic component was softer because it flowed around the urethane particles, apparently other mechanisms were at work (partial curing of the urethane beforehand?) because, in fact, the acrylic is harder than the urethane, as discussed in Section 6.7.2.3. (See especially Figure 6.28.)

At a composition of 50% polyurethane, the polyurethane particles touch, as illustrated in Figure 6.7b. Beyond this composition range, both phases display cocontinuity (Figure 6.7c). Molecular interpenetration is restricted to the interfaces, however. Thus, only the phases interpenetrate in any real sense.

In a separate experiment, Klempner *et al.*[37] hydrolyzed the poly-urethane portion of the IEN in acid. The density of the final product was 0.78 g/cm^3, which can best be interpreted by the presence of extensive void

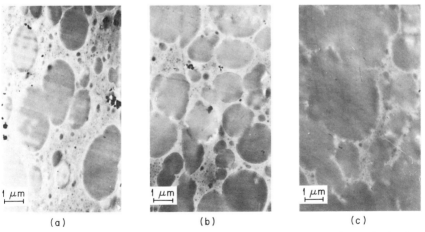

(a)   (b)   (c)

Figure 6.7. Electron micrographs of interpenetrating elastomer networks from mixed poly-acrylate and polyurethane latexes; (a) 70/30; (b) 50/50; (c) 30/70 polyacrylate/polyurethane. Polyurethane stained with osmium tetroxide.[36]

formation. This conclusion is consistent with the morphology in Figure 6.7 and demonstrates that, to a large extent, at least, the two networks were not chemically bound to each other, but rather had to exhibit two continuous phases.

Since the polyacrylate and polyurethane phases have T_g's near 20°C and −30°C, respectively, the IENs were rather soft at room temperature,

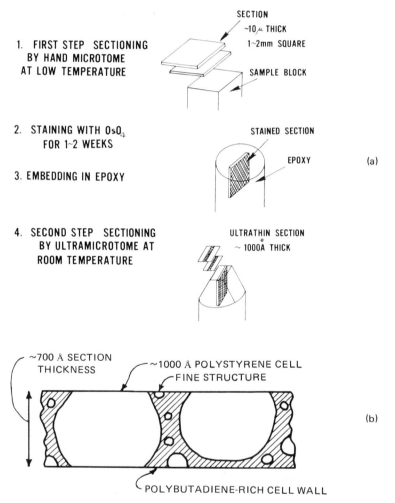

Figure 6.8. The art of sectioning. (a) Schematic explanation of the two-step sectioning method to be used for soft materials.[36] (b) Schematic cross-sectional view of an ultramicrotomed section.

requiring special techniques for sectioning in the ultramicrotome. Matsuo *et al.*[36] used Kato's OsO_4 technique,[38] but had to introduce an additional step for the sectioning of the materials used in their study. In the two-step sectioning method, as schematically described in Figure 6.8, sections as thick as 10 μ were first cut using a hand microtome at liquid nitrogen temperature and then stained with OsO_4 vapor completely to the center of the sections. The stained sections were embedded in an epoxy resin in a transverse way as depicted in Figure 6.8a. They were then submitted to a second step, i.e., ultrathin sectioning at room temperature, the thickness in the latter case being about 1000 Å. A schematic cross-sectional view of an ultramicro-tomed section is shown in Figure 6.8b.[18] Electron mcrographs were taken by the direct observation of the ultrathin sections, leading to Figure 6.7.

6.4.3.2. Sequential Latex IPNs

Although the sequential latex IPNs were long suspected of having a core–shell morphology, it was not until the work of Sionakidis *et al.*[39] that this was established. As described in Section 5.4, a PVC/P(B-*co*-AN) 50/(25–25) latex IPN was prepared. The individual latex particles, shown in Figure 6.9, ranged from 450 to 800 Å. It was considered paramount to establish the actual diameters of the particles, so that the morphological

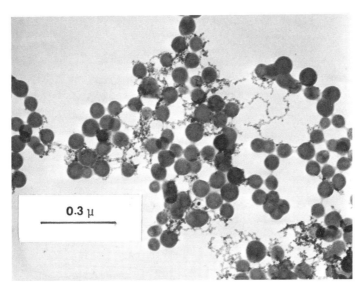

Figure 6.9. PVC/P (B-*co*-AN) finished latex. Electron micrograph of OsO_4-stained particles.[39]

features of the bulk material could positively be related back to the individual particles. The cast and molded products showed more or less identical morphologies (Figure 6.10), with the individual latex particles retaining their individual identities. The major finding is consistent with the notion of a core–shell morphology, where only the shells are stained dark. In the bulk material, of course, the shells form the continuous phase.

A more direct proof of the core–shell morphology itself is available on close inspection of Figure 6.9. Beyond the actual diameters, the key finding relates to the staining intensity within the individual particles.

Three cases can be considered for TEM image contrast of the finished latexes (see Figure 6.11). In the first case, Figure 6.11a, the latex spheres are taken to be homogeneous in composition and to stain uniformly with OsO_4. In TEM, the particles would be darkest in the interior, as expected for the image of any sphere viewed in transmission.

In the second case, Figure 6.11b, the latex spheres are taken to have a stained shell that is sharply demarcated from an unstained core. The transmission image is darkest immediately outside the core, and the core is much brighter than in the homogeneous case.

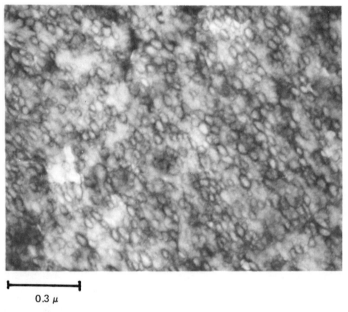

0.3 μ

Figure 6.10. PVC/P(B-*co*-AN) cast film. Electron micrograph of film exposed to OsO_4 for about 45 min after it was sectioned. Film was preheated at 90°C for about 30 min before sectioning.[39]

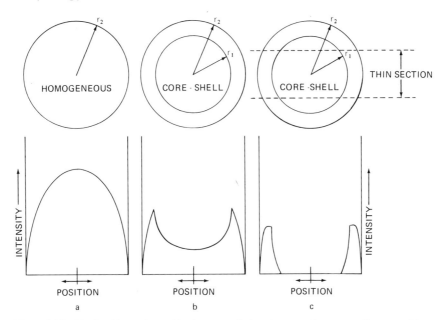

Figure 6.11. Idealized image intensities for transmission electron microscopy of latex particles: (a) homogeneous latex, (b) core–shell latex, (c) ultramicrotomed thin section through core–shell latex. Core/shell ratio, calculated assuming 50 wt % PVC/50 wt % P(B-co-AN) and polymer densities of 1.39 and 1.06 g/cm^3, respectively, gives $r_2 = 1.31r_1$.[39]

For the core–shell limiting case, Figure 6.11c shows that ultramicro-tomed thin sections can exaggerate the image contrast between core and shell. This effect is apparent in Figure 6.10, which shows touching dark rings of the same diameter range as the finished latex particles.

The actual stained latex, Figure 6.9, appears to lie between Figures 6.11a and 6.11b. Most particles are slightly but definitely darker near the exterior than in the center, consistent with a higher concentration of polymer II, P(B-co-AN), toward the exterior. An early stage of latex coalescence is apparent from the attachment points between particles, but these are light in the micrograph because they are only 70 to 150 Å thick.

Several possibilities may be imagined for the detailed morphological structure within the latex particles. First, a smooth gradation of composition from core to shell may exist. Alternatively, but less likely, there may be an abrupt composition boundary between the core and shell. Further, there may be phase separation within the core. Two possibilities are shown in Figure 6.12a.[40] The structure on the right models the known morphology of the ABS latexes.[12] The structure on the left, however, is predicted for two cases: (a) when the latex particles are too small to form cellular

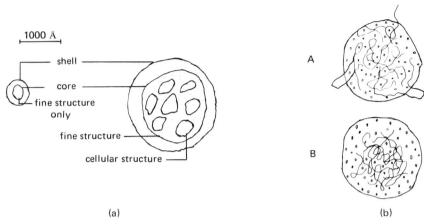

Figure 6.12. Models of latex IPN structure. (a) Model of predicted latex IPN morphology, showing cellular structures, fine structures, and shell–core morphology. (b) Two models of a partly polymerized latex particle. Model A permits some polymer/water contacts, which are "forbidden" in model B.[40]

structures, and (b) when the component polymers are semicompatible. Both of these conditions appear to be true in the present case.

The question of the origin of the core–shell structure, however, permeates the entire field of emulsion polymerization. A brief discussion will shed some light on this complex problem.[40] Referring to Figure 6.12b, polymer loops or ends protruding into the aqueous phase as shown on the left are strongly discouraged or essentially forbidden because of the highly unfavorable heat of mixing that arises upon mixing organic polymer chains and water. Considering the water as part of the "solvent" along with monomer II, let us consider the Flory-type mixing statistics. Inherent in Flory's theory (designed for polymer–solvent systems) is the effect of excluded volume, both intra- and intermolecular exclusions being considered. According to Flory's mixing scheme, whenever a polymer molecule is placed in an unrealistic or impossible position, the entire molecule must be lifted out and placed anew into the medium, each new placement being on a statistical or random basis. In this case, the aqueous phase is essentially forbidden. Thus, the polymer chains already formed (polymer I) will tend to be concentrated in the central portion of each latex particle so as to avoid polymer–water contacts. Monomer II occupies the remaining sites, mostly near the surface but some within the interior as shown in Figure 6.12b, right, and on polymerization forms a more or less permanent shell. (This model, as such, does not distinguish between the monomer and polymer being the same or different.)

6.4.4. Effect of Composition Ratio

6.4.4.1. Sequential IPNs

Returning to sequential IPNs, at low percentages of polymer II one intuitively expects the domain size of polymer II to increase with the quantity of that polymer incorporated. For midrange and higher concentrations, however, the relationship becomes more complicated.

Yenwo et al.[41,42] studied the variation of the domain size with both crosslink level (of polymer I) and overall composition (see Figure 6.13). The castor oil–urethane/polystyrene IPNs shown were prepared by sequential polymerization.

The crosslink density of network I is determined by the NCO/OH ratio. Theoretically, the minimum value to form a continuous network is near

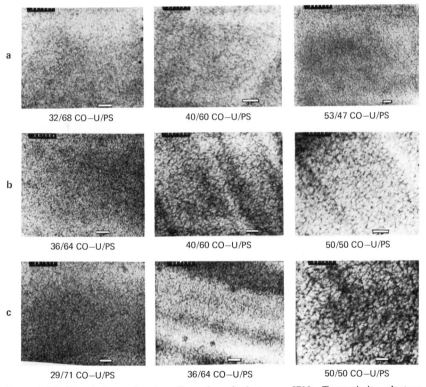

Figure 6.13. Morphology of castor oil–urethane/polystyrene IPNs. Transmission electron microscopy, osmium tetroxide staining of the castor oil component.[41] (a) NCO/OH = 0.95; (b) NCO/OH = 0.85; (c) NCO/OH = 0.75. The scale indicator at the bottom of each photograph is 0.1 μm.

0.67, since castor oil is trifunctional. The maximum crosslink level, of course, is attained at a value of 1.0. As the crosslink level increases (bottom to top), the phase domain sizes in Figure 6.13 are seen to decrease. This is in line with the preceding discussion.

As the quantity of polystyrene in the IPN is increased through the midrange composition as shown in Figure 6.13, the phase domain sizes are seen to decrease (right to left). As will be seen below, changes in this direction can be predicted theoretically. From a structural point of view, a higher level of polymer II causes an increase in the number of domains that are forming.

The total effect remains unclear, however. For sequential IPNs based on PEA/PS, Huelck et al.[17] found slight increases in domain size as the composition ranged from 75/25 to 25/75. Unfortunately, at the latter composition the cellular structure was somewhat broken down and the dimensions uncertain.

6.4.4.2. SINs

Similar changes were observed by Allen et al.[22,43] as their PU/PMMA semi-SINs were made with larger amounts of PMMA. Here, however, the effect was attributed, in part, to the looser PU network that formed in the increasingly dilute urethane prepolymer solutions.

For PU/PMMA SINs with similar rates of polymerization, Kim et al.[27] showed that phase inversion occurs in the composition range of 20%–40% PMMA. When either component exceeds 85%, however, the morphology becomes more complicated. This work and that of Touhsaent et al.[34,35] both indicate that phase domain size and continuity depend sensitively on the relative reaction rates for the two polymers, as well as on composition ratio.

6.5. QUANTITATIVE EXPRESSIONS FOR PHASE DOMAIN SIZE

6.5.1. Block Copolymers and ABCPs

Meier[20,26] considered the morphologies of diblock copolymers and found that for dispersed spheres

$$R = k\alpha CM^{1/2} \tag{6.1}$$

where R represents the domain radius, k is a constant characteristic of the morphology (1.33 for spheres), C represents an experimental constant relating the unperturbed root-mean-square end-to-end distance of a chain to the square root of its molecular weight M, and α is the ratio of the perturbed to unperturbed chain dimensions. From electron microscopy studies on block copolymers, this equation was confirmed.

Equation (6.1) can be applied to other systems where the molecular weight controls domain sizes. For ABCPs, Eastmond and Smith[25] found

$$R = 0.06M_n^{1/2} \tag{6.2}$$

for the domain sizes of the B blocks.

In both the block copolymer case and the AB crosslinked copolymer case, the phase domain dimensions depend upon the molecular weight of the polymer in the same domain. Use of equation (6.1) above, employing either polymer I or polymer II, leads to an altogether incorrect analysis for IPNs. In the case of IPNs, presented below, the phase domain sizes of polymer II are shown to depend upon the molecular weight between crosslinks $(\bar{M}_c)$ of polymer I.

6.5.2. Sequential IPNs and Semi-IPNs

Donatelli *et al.*[21] derived a semiempirical equation for the phase domain size of semi-IPNs of the first kind. Under certain conditions, the equation applies to full IPNs. Thermodynamically based, the equation has as its principal variables the crosslink density of polymer I, the mass fraction of polymer II, and the interfacial tension.

In the derivation presented below, the IPN morphology is imagined to evolve over the following hypothetical path, and the free-energy change over this route is developed. Initially, in state 1, polymer I and polymer II are completely separated. At an intermediate level, state 2, polymer I network is uniformly swollen with polymer II, and the two polymers form a randomly mixed state. Finally, in state 3, polymer II is allowed to phase separate into spherical domains within the polymer I network, with no change in the volume of the system. It is assumed that, on a macroscopic scale, phase separation does not alter the state of swelling of the polymer I network, or the total volume of the system. It is also presumed that thermodynamic equilibrium is achieved in each step, which represents a significant approximation in the case of IPNs.

The free-energy change for polymer II domain formation, ΔG_d, is the sum of the free-energy changes from states 1 to 2 and from states 2 to 3:

$$\Delta G_d = \Delta G_{12} + \Delta G_{23} \tag{6.3}$$

The free-energy change from state 1 to state 2 is equal to the sum of the ordinary free energy of mixing, ΔG_m, and the elastic free-energy change of the polymer I network which is uniformly swollen with polymer II, ΔG_{el}:

$$\Delta G_{12} = \Delta G_m + \Delta G_{el} \qquad (6.4)$$

The free-energy change from state 2 to state 3 is equal to the sum of the ordinary free energy of demixing, ΔG_{dm}, and the surface free energy for domain formation, ΔG_s:

$$\Delta G_{23} = \Delta G_{dm} + \Delta G_s \qquad (6.5)$$

6.5.2.1. Free Energy of Mixing

The term ΔG_m in equation (6.4) is represented by equation (6.6), where ΔH_m and ΔS_m represent the heat and entropy of mixing, respectively, and T represents the absolute temperature at which mixing occurs:

$$\Delta G_m = \Delta H_m - T\Delta S_m \qquad (6.6)$$

The heat of mixing of nonpolar substances is given by a Van Laar–Hildenbrand–Scatchard expression:[44]

$$\Delta H_m = V(\delta_1 - \delta_2)^2 v_1 v_2 \qquad (6.7)$$

where v_1 and v_2 are the volume fraction of polymers 1 and 2, respectively; δ_1 and δ_2 are solubility parameters of polymers 1 and 2, respectively; and V is the total volume of mixture.

Using a lattice model, Flory[45] has developed an expression for the entropy of mixing between a solvent and polymer. Extending this development to the case of a mixture of two polymers, the entropy change can be approximated by

$$\Delta S_m = -k(N_1 \ln v_1 + N_2 \ln v_2) \qquad (6.8)$$

where N_1, N_2 are the numbers of polymer 1 and 2 molecules, respectively, and k is Boltzmann's constant.

The ordinary free energy of mixing can now be expressed as follows [see equation (2.2)]:

$$\Delta G_m = V(\delta_1 - \delta_2)^2 v_1 v_2 + kT(N_1 \ln v_1 + N_2 \ln v_2) \qquad (6.9)$$

Equation (6.9) and below are conceived on a "per unit volume" basis.

6.5.2.2. Elastic Free-Energy Change

The elastic free-energy change is primarily an entropy contribution arising from the uniform swelling of the polymer I network by polymer II.

This can be expressed as

$$\Delta G_{el} = -T\Delta S_{el} \tag{6.10}$$

Flory has derived the entropy change for the isotropic swelling of a rubber network:

$$\Delta S_{el} = -(R\nu_1/2)(3\alpha_1^2 - 3 - \ln \alpha_1^3) \tag{6.11}$$

where ν_1 is the effective number of moles of polymer chains in network I, and α_1 is the ratio of perturbed to unperturbed chain dimensions for network I.

The chain expansion parameter, α_1, can be expressed as

$$\alpha_1 = (r/r_0)_1 \tag{6.12}$$

where r_0 and r are the root-mean-square (rms) end-to-end distances of the unperturbed and perturbed molecule, respectively, for polymer I. Also, r_0, is equal to

$$r_0 = KM^{1/2} \tag{6.13}$$

where K is a known constant for most polymers and M is the average molecular weight of the polymer. Combining equations (6.12) and (6.13) and substituting into equation (6.11) with the assumptions that r equals r_1, the rms distance between crosslink sites in polymer network I, and M equals M_{c1}, the number-average molecular weight between crosslink sites for the polymer I network, yields

$$\Delta S_{el} = -\left[\frac{k\nu_1}{2}\right]\left[\frac{3r_1^2}{K^2 M_{c1}} - 3 - \ln\left(\frac{r_1}{KM_{c1}^{1/2}}\right)^3\right] \tag{6.14}$$

The elastic free-energy change from state 1 to state 2 is then

$$\Delta G_{el} = \left[\frac{kT\nu_1}{2}\right]\left[\frac{3r_1^2}{K^2 M_{c1}} - 3 - \ln\left(\frac{r_1}{KM_{c1}^{1/2}}\right)^3\right] \tag{6.15}$$

6.5.2.3. Free Energy of Demixing

Returning to equation (6.5), ΔG_{dm} is equal to

$$\Delta G_{dm} = \Delta H_{dm} - T\Delta S_{dm} \tag{6.16}$$

where ΔH_{dm} and ΔS_{dm} represent the heat and entropy of demixing, respectively. The heat of demixing can be approximated by a negative heat of mixing which is the opposite of equation (6.7):

$$\Delta H_{dm} = -\Delta H_m = -V(\delta_1 - \delta_2)^2 v_1 v_2 \tag{6.17}$$

The entropy of demixing is the entropy change between states 2 and 3. At state 2, ΔS is equal to equation (6.8), or

$$\Delta S_2 = -k(N_1 \ln v_1 + N_2 \ln v_2) \tag{6.18}$$

From statistical thermodynamics, the entropy of a system in a specified state can be defined in terms of the number of possible arrangements of the molecules composing the system which are consonant with the state of the system.[46] Each possible arrangement is called a complexion of the system, and the entropy is defined by equation (6.19), first developed by Boltzmann, where Ω equals the number of complexions:

$$\Delta S = k \ln \Omega \tag{6.19}$$

In the final state, the two polymers are considered to be completely phase separated. Polymer I forms the continuous phase, and polymer II forms the discontinuous phase. N_d domains of polymer II are formed with a total of N_2 polymer II molecules occupying the domains. If a particular polymer II molecule is able to enter any of the N_d domains during phase separation, the number of complexions is

$$\Omega = (N_d)^{N_2} \tag{6.20}$$

For state 3, ΔS_3 can now be expressed as

$$\Delta S_3 = k \ln (N_d)^{N_2} = kN_2 \ln N_d \tag{6.21}$$

(The entropy of polymer I, a pure phase, may be considered zero.) By subtracting equation (6.18) from equation (6.21), the entropy of demixing is obtained:

$$\Delta S_{dm} = \Delta S_3 - \Delta S_2 = k(N_2 \ln N_d + N_1 \ln v_1 + N_2 \ln v_2) \tag{6.22}$$

Substituting equations (6.17) and (6.22) into equation (6.16) yields the ordinary free energy of demixing:

$$\Delta G_{dm} = -V(\delta_1 - \delta_2)^2 v_1 v_2 - kT(N_2 \ln N_d + N_1 \ln v_1 + N_2 \ln v_2) \tag{6.23}$$

6.5.2.4. Surface Free Energy of Domain Formation

A special problem to be considered involves interfacial free energy. The surface free energy for polymer II domain formation is equal to

$$\Delta G_s = \gamma A_s \tag{6.24}$$

where γ is the interfacial energy and A_s is the surface area of interaction between polymer I and polymer II. If the polymer II domains are assumed to be spherical, the surface area is equal to

$$A_s = N_d \pi D_2^2 \tag{6.25}$$

where D_2 is the average polymer II domain diameter. Substituting equation (6.25) into (6.24) yields

$$\Delta G_s = \pi \gamma N_d D_2^2 \tag{6.26}$$

6.5.2.5. Free-Energy Change of Domain Formation

The free-energy change for polymer II domain formation is obtained by adding equations (6.9), (6.15), (6.23), and (6.26):

$$\Delta G_d = \pi \gamma N_d D_2^2 - kTN_2 \ln N_d + \left[\frac{kT\nu_1}{2}\right]\left[\frac{3r_1^2}{K^2 M_{c1}} - 3 - \ln\left(\frac{r_1}{KM_{c1}^{1/2}}\right)^3\right] \tag{6.27}$$

Since all the polymer II is assumed to be located in the domains, the volume of polymer II is

$$V_2 = N_d \pi D_2^3/6 \tag{6.28}$$

and the number of polymer II domains is

$$N_d = \frac{6V_2}{\pi D_2^3} \tag{6.29}$$

From Figures 2.3 and 6.13, the polymer II domain size is clearly a function of the degree of crosslinking of the polymer I network.

For simplicity, it is assumed that the domain size of polymer II is a linear function of the rms end-to-end distance between crosslink sites of the polymer I network, expressed by equation (6.30). The quantity C is a proportionality constant relating the physical distance between the crosslink sites of the polymer I and the domain dimensions of polymer II (see Figure 6.14):

$$D_2 = Cr_1 \tag{6.30}$$

Equation (6.27) now can be rewritten in the form

$$\Delta G_d = \frac{6\gamma V_2}{D_2} - RT\bar{N}_2 \ln\left(\frac{6V_2}{\pi D_2^3}\right)$$
$$+ \left[\frac{RT\nu_1}{2}\right]\left[\frac{3D_2^2}{C^2 K^2 M_{c1}} - 3 - \ln\left(\frac{D_2}{CKM_{c1}^{1/2}}\right)^3\right] \tag{6.31}$$

where $\bar{N}_2$ is the number of moles of polymer II and ν_1 is the effective number of moles of crosslinked chains in the polymer I network.

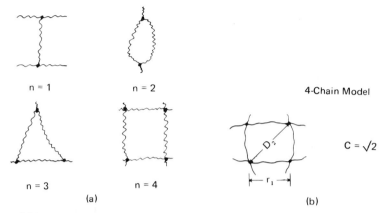

Figure 6.14. Molecular models for evaluating C. (a) Schematic diagram of some simple crosslink clusters, useful for estimating the parameter n, and hence C, in equation (6.36). (b) The four-chain model, showing the relationship between r_1 and D_2.[22]

In order to determine the domain diameter which gives the minimum free-energy change, the first derivative of equation (6.31), $d(\Delta G_d)/dD_2$, is equated to zero and solved implicitly for D_2. This procedure yields

$$\frac{\nu_1 D_2^3}{C^2 K^2 M_{c1}} + \left(\bar{N}_2 - \frac{\nu_1}{2}\right)D_2 = \frac{2\gamma V_2}{RT} \qquad (6.32)$$

The number of moles of polymer II is equal to

$$\bar{N}_2 = W_2/M_2 \qquad (6.33)$$

where W_2 and M_2 are the weight fraction and molecular weight, respectively. The number-average molecular weight between crosslink sites for polymer I, M_{c1}, can be related to the crosslink density ν_1 by the following equation[45]:

$$\frac{1}{M_{c1}} = \frac{\nu_1 \bar{v}_1}{V_1} + \frac{2}{M_1} \qquad (6.34)$$

where $V_1 \bar{v}_1$, and M_1 are the volume, specific volume, and primary molecular weight, respectively, of polymer I. The last term on the right-hand side of equation (6.34) arises from the need to correct for dangling end chains in lightly crosslinked materials. Assuming a density of 1.0 g/cm^3 for both polymers so that the weights are equivalent to the volumes along with the relationship (on a per unit volume basis)

$$W_1 + W_2 = 1 \qquad (6.35)$$

and substituting equations (6.33)–(6.35) into equation (6.32), the equation for the polymer II domain diameter is obtained:

$$\left(\frac{\nu_1}{C^2 K^2}\right)\left(\frac{\nu_1}{1-W_2}+\frac{2}{M_1}\right)D_2^3 + \left(\frac{W_2}{M_2}+\frac{\nu_1}{2}\right)D_2 = \frac{2\gamma W_2}{RT} \qquad (6.36)$$

Finally, the constant C from equation (6.30) is evaluated in the following manner. Several models may be used to evaluate the constant C, where the domains are pictured as surrounded by a crosslink cluster containing n chains (see Figure 6.14). The $n = 1$ model predicts $C = 1$. The four-chain model predicts $C = \sqrt{2}^{1/2}$. In reality, all possible chain combinations leading back to the point of origin have to be considered, and a weighted sum from $n = 1$ to $n = \infty$ should be used. For simplicity, n will be assumed to be 4, yielding $C = \sqrt{2}^{1/2}$, which assumes that an average of four polymer I chain segments are required to circumnavigate an average polymer II domain.

Equation (6.36) can be simplified from a cubic form to a linear form through appropriate algebraic substitutions for C and D. The reduced form,

$$D_2 = \frac{2\gamma W_2}{RT\nu_1\{[1/(1-W_2)]^{2/3} - \frac{1}{2}\}} \qquad (6.36')$$

shows that D_2 is directly proportional to the interfacial energy, and inversely proportional to the crosslink density of network I. The quantity D_2 approaches zero as W_2 approaches zero.*

6.5.2.6. Applications

The case for the full IPN (both polymers crosslinked) can be approximated by taking $M_2 = \infty$. Equation (6.36) has been applied to four systems to date:

1. SBR/PS (see Figure 2.3), where the crosslink density of polymer I was the principal variable. The results are shown in Table 6.2.[21]

2. Castor oil–urethane/PS IPNs, where both composition and crosslink level were varied.[47] The results are summarized in Table 6.3. In both

---

* In equations 6.36 and 6.36', ν_1 represents the crosslink density of network I per unit volume of total IPN. J. Michel, S. C. Hargest, and L. H. Sperling, *J. Appl. Polym. Sci.* (to be published), proposed a form where ν_1 represents the crosslink density of the dry network I, and ϕ_1 and ϕ_2 represent the volume fractions of networks I and II, respectively.

$$D_2 = \frac{2\gamma\phi_2}{RT\nu_1\phi_1[(1/\phi_1)^{2/3} - 1/2]} \qquad (6.36'')$$

Table 6.2. Experimental and Theoretical PS Domain Sizes in IPNs and Semi-1 Compositions, $\gamma = 3$ dyn/cm[21]

Type	Composition SBR/PS(%)	Experimental PS domain size (Å)	Theory (Å)
(Low)[a] semi-1	20/80	1500	1250
(High) semi-1	18/82	600	500
(Low) IPN	22/78	1100	1300
(High) IPN	21/79	650	480

[a] Low refers to 0.1% crosslinking agent in the SBR; "high" refers to 0.2%.

Tables 6.2 and 6.3, the comparison between theory and experiment is better than warranted from the semiempirical nature of the equation.

3. PEA/PMMA IPNs, see Figure 6.2a, where, as a first approximation, the interfacial tension, γ, may be assumed to be zero. Domains of the order of 60–100 Å are predicted. This modifies earlier conclusions of semicompatibility for this system, suggesting that fine structure is always to be expected in sequential IPNs. Physically, this might result from the concentration of monomer II in small regions of space (within polymer I) that statistically have lower than average crosslinking levels.

4. The PS/PS homo-IPNs, discussed in Chapter 4, also show evidence of domain formation even though both polymers are identical. In this case, γ is certainly zero and the predicted 60–100 Å domain sizes are indeed found.[48]

Table 6.3. Polystyrene Domain Sizes in Castor Oil–Urethane IPNs[47]

IPN sample	NCO/OH ratio	Weight fraction PS	Polystyrene domain size (Å)	
			Experimental[a]	Theoretical[b]
1	0.95	0.68	250	323
2	0.95	0.60	300	325
3	0.95	0.47	370	327
4	0.85	0.64	300	344
5	0.85	0.60	350	345
6	0.85	0.50	430	347
7	0.75	0.71	350	432
8	0.75	0.64	410	443
9	0.75	0.50	550	452

[a] Estimated from electron microscopy.
[b] Theoretical, from equation (6.36).

6.6. INDUCED MORPHOLOGICAL CHANGES

6.6.1. Inverse IPNs

Before leaving off the discussion of morphology, two additional structures must be presented in order to appreciate the range of materials that can be prepared. In Figure 6.2b a poly (ethyl acrylate)/polystyrene IPN was shown, polymerized in that order. Its topological designation may be written

$$C_1 \, O_I \, C_2 \qquad (6.37)$$

Inverting the order of synthesis, i.e., PS first, yields

$$C_2 \, O_I \, C_1 \qquad (6.38)$$

Assuming that the stained polymer (PEA) is the same, a positive–negative photographic inversion may be expected. Figure 6.15[17] presents a PS/PEA inverse IPN, which should be compared with Figure 6.2b. While a general inversion of phase structure is apparent, the cell walls of the inverse composition appear to be thicker and the fine structure at 100 Å level appears to be within the cells, rather than in the cell walls.

6.6.2. Decrosslinked IPNs

The complexity of an IPN allows yet another morphology to be developed, through decrosslinking.[49] Specific decrosslinking without

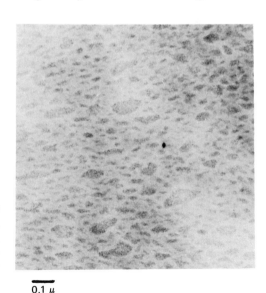

Figure 6.15. Electron micrograph of IPN 12: 50.7 PS/49.3 PEAB. This structure is the inverse of that shown in Figure 6.2b. The first component forms the matrix material and exhibits greater continuity than the second component.[17]

0.1 μ

degradation of the primary chains can be accomplished through the use of acrylic acid anhydride, AAA, as a crosslinker. In order to facilitate comparisons, PEA/PS materials much like that shown in Figure 6.2b were prepared, except that AAA was used as the crosslinker instead of TEGDM or DVB. The change in the topology can be represented

$$\beta(C_1 \, O_I \, C_2) \to P_1 \, O_M \, P_2 \tag{6.39}$$

where β is a function which operates on $(C_1 \, O_I \, C_2)$, transforming it to a chemical blend. The electron micrograph in Figure 6.16 represents a decrosslinked and annealed IPN sample, while that shown in Figure 6.17 represents a decrosslinked and annealed semi-I material. In particular, the PS in Figure 6.17 now appears to have gained a degree of continuity not present in Figure 6.2b.

Some of the changes in annealing can be explained by assuming that the polymer chains are not in the same state in the IPN as in the homopolymer. In the case of an IPN, polymer I is strained because it has been swollen with monomer II during synthesis. Polymer II develops unstrained during synthesis (see Chapter 4). During annealing (after decrosslinking), the domain structure is rearranged to minimize the molecular strain, yielding a minimum in the free energy.

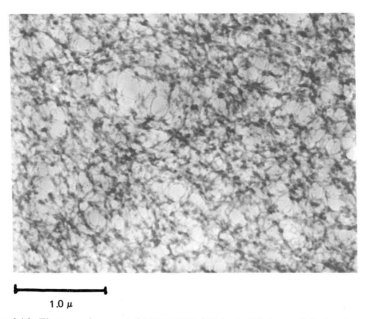

1.0 μ

Figure 6.16. Electron micrograph 28,000 × PEA/PS (code 13) decrosslinked and annealed. Note the appearance of oriented or fibrillar PEA structures.[49]

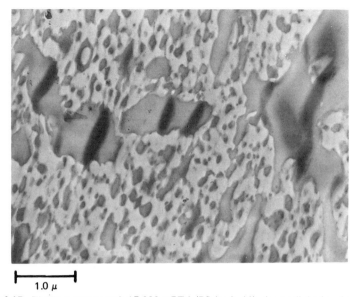

I———I
  1.0 μ

Figure 6.17. Electron micrograph 17,000 × PEA/PS (code 11), decrosslinked and annealed. While the PS now assumes a degree of continuity, the PEA has both continuous and discontinuous portions.[49]

Further electron micrographs will be found elsewhere in this book, particularly in Chapter 2. Certainly not all of the possible IPN morphologies have been discovered yet. While equation (6.36) describes the phase domain size in sequential IPNs, the correlation between synthetic detail and morphology is still in its infancy. Nonetheless, the experimentally known morphologies provide an important link between synthesis and properties.

6.7. PHYSICAL AND GLASS TRANSITION BEHAVIOR

6.7.1. General Properties

An IPN can be distinguished from simple polymer blends, blocks, and grafts because (1) an IPN swells, but does not dissolve in simple solvents, and (2) creep and flow are suppressed. The behavior of an IPN depends on the properties of the components, the phase morphology, and the interactions between the phases.

On a more subtle level, the deviations in density from those determined by the weight fractions of the components affect the glass transition temperature, T_g, of the material; greater than average densities presumably lead to higher T_g's. The evidence seems divided. The first studies were done on homo-IPNs, see Chapter 4. Millar observed very slight increases in density in his polystyrene/polystyrene IPNs[50] which could have been caused by a filling of small quantities of supermolecular holes. However, Shibayama and Suzuki[51,52] observed no difference in density between PS/PS IPNs and their constituent networks. Klempner et al.[53] also observed no difference in density in their polyurethane/polyacrylate latex IPNs. Detailed data showing only additive effects are provided by Lipatov et al. on an Aerosil-filled polyurethane/polystyrene sequential IPN (see Table 6.4)[54] The theoretical density values, ρ_{theor}, in Table 6.4 were calculated on the basis of an additivity scheme.

In contrast to the Lipatov et al.[52] data, Kim et al.[55] found that SINs prepared from polyurethane and polystyrene did show a densification, as illustrated in Figure 6.18. This densification was correlated with shifts in the T_g.

However, the data in the same figure on polyurethane/poly(methyl methacrylate) SINs show strict additivity. While the evidence indicates that important densifications sometimes take place, it is interesting to note that no investigator has suggested that a decrease in density was probable. The

Table 6.4. Density of Filled Polyurethane/Polystyrene IPNs[54]

Filler, %	ρ for network I PU g/cm³	ρ for network II PS g/cm³ 2%ᵃ DVB	3% DVB	IPN 1% DVB w_2/w_1	ρ_{exp} g/cm³	ρ_{theor} g/cm³	2% DVB w_2/w_1	ρ_{exp} g/cm³	ρ_{theor}	3% DVB w_2/w_1	ρ_{exp} g/cm³	ρ_{theor}
0	1.244	1.054	1.053	0.505	1.196	1.193	0.283	1.195	1.196	0.289	1.190	1.195
				0.359	1.188	1.188	0.286	1.194	1.198	0.308	1.194	1.193
0.5	1.245	1.056	1.058	0.412	1.184	1.184	0.175	1.206	1.215	0.242	1.202	1.205
				0.477	1.777	1.178	0.256	1.199	1.203	0.240	1.202	1.205
				—	—	—	0.305	1.997	1.197	0.297	1.196	1.197
				—	—	—	0.449	1.199	1.176	0.50	1.196	1.175
1.0	1.246	1.057	1.060	0.492	1.182	1.179	0.289	1.198	1.202	0.316	1.189	1.192
				0.554	1.174	1.173	0.462	1.197	1.183	0.328	1.195	1.196
				—	—	—	—	—	—	0.666	1.165	1.164
				—	—	—	—	—	—	2.985	1.137	1.111
5.0	1.267	1.070	1.080	0.425	1.199	1.206	0.149	1.219	1.248	0.140	1.208	1.252
				0.498	1.199	1.198	0.443	1.201	1.210	0.697	1.207	1.179

ᵃ Divinyl benzene as crosslinker for the PS network.
ᵇ Experimental densities determined by hydrostatic weighing in isooctane.

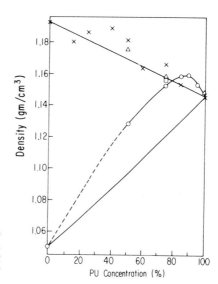

Figure 6.18. Density versus polyurethane concentration (straight lines are based on volume additivity: (×) UCMC IPNs; (○) UCSC IPNs; (△) linear blends; (□) pseudo-IPNs.[55]

extremely interesting and related problem of possible increases in crosslink density through interpenetration will be relegated to a later section; also see Chapter 4.

Another general property of two-phased materials is their optical appearance. Because the phase domains usually have different refractive indices, they scatter light,* which results in a hazy, translucent, or milk-white appearance. Such a clouded appearance, along with two glass transition temperatures, has along been accepted as primary evidence of phase separation. Typical data are presented in Table 6.5[54] for a fluorocarbon elastomer/acrylic semi-II IPN system. The turbidity, τ, is based on Beer's law:

$$I = I_0 \, e^{-\tau x} \tag{6.40}$$

---

* In general, the quantity of scattered light, and hence the opacity, depends upon two parameters:

1. The amount of scattered light increases as the square of the difference in refractive index of the two phases. Thus, polymer blends, grafts, and IPNs will be clear if the refractive indexes nearly match.

2. For particles small compared to the wavelength of the light, scattering increases as the sixth power of the particle diameter. For domains near the wavelength in size, scattering increases as the square of the wavelength. This latter accounts for the blue–white opalescence frequently seen in IPNs and graft copolymers. For objects large compared to the wavelength of light (tables, chairs, etc.) the wavelength dependence on scattering is effectively zero and the phenomenon is commonly called reflection.

Table 6.5. Turbidity of Fluorocarbon Semi-II IPNs[56]

| % TMPTM[a] | Crosslinking radiation | | τ |
	Method	Dose (Mrad)	
0.0	—	0.0	1.10
3.0	β	5.0	0.15
3.0	γ	5.2	0.26
3.0	uv	—	0.46

[a] Trimethylolpropane trimethacrylate.

where I_0 is the initial light intensity ($\lambda = 460 \ \mu$m was employed), I is the transmitted light intensity, and x represents the sample thickness. A correction of 4% reflection from each surface was made. The results in Table 6.5 illustrate the grafting effect of high-energy β or γ irradiation vs. uv irradiation. The former two suppress domain formation through grafting resulting in a clearer specimen than the latter. By visual inspection, samples 3–4 mm thick are nearly window glass clear below turbidities of 0.15, but are noticeably cloudy above $\tau = 0.40$. At that thickness, the material becomes opaque near $\tau = 1.00$.

At the same thickness (3–4 mm), the materials in Figure 6.1a were substantially clear. The materials in Figure 6.2b were translucent, and those in Figure 2.3 were opaque. The commercial HiPS and ABS plastics are usually opaque white.

6.7.2. Glass Transition Behavior

While electron microscopy reveals information about domain size, shape, and continuity, in general only slight hints relevant to polymer composition within the phases can be obtained. Fortunately, the study of glass transitions is complementary to morphological studies.

Through an analysis of the shifts and broadenings of the T_g's of the component polymers, quantitative information can be extracted about the composition of the phases. Since analytical studies of the thermodynamics of mixing, including such concepts as the approach to equilibrium, calculation of the free energy of mixing, etc.[57–60] are coming to the fore, even older data may be enhanced by a reevaluation from a different perspective.

The T_g's of polyblends in general, and of IPNs in particular, can be studied in a variety of ways. Unfortunately, the study of IPNs shares a problem common to all of the phase-separated blends, grafts, and blocks, and indeed to virtually all science: no one investigator has systematically studied the same series of samples by all of the important techniques.

6.7.2.1. Theoretical Aspects

Two well-known copolymer equations may be adopted in calculating the T_g's of compatible IPNs and SINs:

$$\frac{1}{T_g} = \frac{w_1}{T_{g1}} + \frac{w_2}{T_{g2}} \qquad (6.41)$$

$$T_g = w_1 T_{g1} + w_2 T_{g2} \qquad (6.42)$$

where T_{g1} and T_{g2} represent the glass transition temperatures of polymer I and polymer II, respectively, w_1 and w_2 are their weight fractions, and T_g is the predicted value of the glass transition temperature of the IPN or SIN. (Alternatively, T_g may be considered the glass temperature of a homogeneous phase in a phase-separated material. See Section 6.7.2.4.)

However, the Frisch team carried the analysis further.[61,62] They began with the concept that chemical cross-linking in conventional polymers raises T_g. If T_g is the glass temperature of the crosslinked polymer, T_{g0} the glass temperature of the uncrosslinked polymer, X_c the mole fraction of monomer units which are crosslinked in the polymer, $\varepsilon_x/\varepsilon_m$ the ratio of the lattice energies for crosslinked and uncrosslinked polymer and F_x/F_m the ratio of segmental mobilities for the same two polymers, then the DiBenedetto equation[63] may be written

$$\frac{T_g - T_{g0}}{T_{g0}} = \frac{[\varepsilon_x/\varepsilon_m - F_x/F_m]X_c}{1 - (1 - F_x/F_m)X_c} \qquad (6.43)$$

Any copolymer effect due to crosslinking is to be accounted for by modifying T_{g0}. For chemically crosslinked polymers $\varepsilon_x/\varepsilon_m \neq 1$ (DiBenedetto estimated this to be about 1.2) and the mobility of a chemically crosslinked segment $F_x \ll F_m$ so that $F_x/F_m \simeq 0$. Hence, equation (6.43) simplifies in first approximation to

$$\frac{T_g - T_{g0}}{T_{g0}} \simeq \frac{1.2X_c}{1 - X_c} \qquad (6.44)$$

which exhibits the often observed experimental increase of T_g with X_c.

Equation (6.43) may be transformed in such a way as to determine the increase in physical crosslinks, if any. (See Chapter 4 for examples of alternate treatments.) The quantity X_c' will now represent increases in physical crosslink density caused by interpenetration in an IPN.

In the case of a compatible SIN or IPN, equation (6.43) must be modified by replacing T_{g0} with $(T_g)_{av}$. In the first approximation, this should account for the copolymer effect which is obviously present in mutually compatible SINs or IPNs. Next it is noted that $\varepsilon_x/\varepsilon_m \simeq 1$, since the mer units of both networks are not extensively modified on forming an SIN from preexisting homopolymer networks. Thus equation (6.43) now becomes

$$\frac{T_g - (T_g)_{av}}{(T_g)_{av}} = \frac{[1 - F_x/F_m]X_c'}{1 - (1 - F_x/F_m)X_c'} \tag{6.45}$$

with X_c' the entanglement mole fraction. If F_x/F_m is still much smaller than unity, see above, equation (6.45) may be written

$$\frac{T_g - (T_g)_{av}}{(T_g)_{av}} = \frac{X_c'}{1 - X_c'} \tag{6.46}$$

Values of $T_g - (T_g)_{av}$, or better, values of $T_g - T_g$ (calc), where T_g (calc) and $(T_g)_{av}$ are derived from equations (6.41) and (6.42), respectively, may be used to estimate the increase (or decrease!) in physical crosslink densities due to interpenetration.

It should be pointed out, however, that the crosslink level must be fairly high for equation (6.44) to yield accurate results. Likewise, the increase in physical crosslink level on interpenetration would have to be very significant before equation (6.46) could yield important results. Neither equation is sensitive to small changes. However, an alternate explanation may be more correct if values of X_c' are negative.

In general, secondary intramolecular bonding of a network, Van der Waals, or hydrogen bonding (if present) are reduced by the permanent entanglement of portions of two different networks; in those cases the mobilities of the segments of an IPN or SIN, F_x, may be larger than in the noninterpenetrating separate network, F_m, i.e., $F_x/F_m > 1$. Setting the quantity θ equal to

$$\theta = [(F_x/F_m) - 1]X_c' > 0 \tag{6.47}$$

$$1 \geq X_c' \geq 0 \tag{6.48}$$

equation (6.45) can be rewritten as

$$\frac{T_g - (T_g)_{av}}{(T_g)_{av}} = -\frac{\theta}{1 + \theta} \tag{6.49}$$

This predicts that the T_g of SIN or IPN would be less than or equal to $(T_g)_{av}$, the relative negative shift being given quantitatively by $\theta/(1 + \theta)$, which increases monotonically from zero to $(F_x - F_m)/F_x$ as X_c' goes from zero to unity.

It must be emphasized that X_c' is a measure of the extra crosslink density due to interpenetration in the above derivation. This interesting, and controversial, issue constitutes the fulcrum of discussion about the very nature of IPNs, and formed the basis for Chapter 4. Whether or not the T_g is raised because of the postulated increase in physical crosslink level, the quantity θ forms a useful measure of nonideality, however.

6.7.2.2. Nonmechanical Methods

The methods available to measure T_g may be conveniently divided into two groups: mechanical and nonmechanical. Perhaps the most widely used nonmechanical method is differential scanning calorimetry, DSC.

Frisch et al.[61] studied a SIN prepared from a polyurethane and a polyacrylate (see Figure 6.19). Each of the SINs showed only one T_g, intermediate in temperature to the T_g's of the component networks. Together with the noted clarity of the films, Frisch et al. concluded that the materials were thermodynamically compatible. The most likely reason for the complete mutual solubility is extensive grafting (see Section 5.5). The methacrylic acid and hydroxyethyl methacrylate probably react to greater or lesser extents with the MDI component. Numerical values for the PU/PA SINs are found in Table 6.6.[61]

Equation (6.41) yields the calculated value of T_g in Table 6.6, while equation (6.42) yields the average T_g. In each case, the experimental T_g lies between the two T_g's obtained from equations (6.41) and (6.42). Values of the quantity for the SINs are shown in Table 6.6. (Note that numerically,

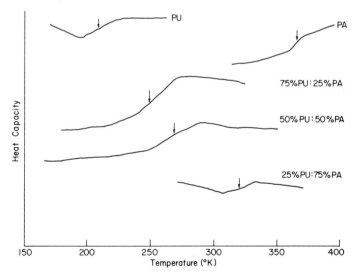

Figure 6.19. Heat capacity, vs. temperature for polyurethane/polyacrylate SINs. Glass transition temperatures are indicated.[61]

Table 6.6. Glass Transition Temperatures of Polyurethane/Polyacrylate SINs[61]

Composition, percent	T_g, °K (expt.)	T_g, °K (calc.)[a]	T_g, °K (av.)[b]	θ^c
100 percent PU	209	—	—	—
75 percent PU/25 percent PA	246	234	249	0.0122
50 percent PU/50 percent PA	274	267	288	0.0512
25 percent PU/75 percent PA	321	317	327	0.0187
100 percent PA	367	—	—	—

[a] Calculated: $1/T_g = W_1/T_{g1} + W_2/T_{g2}$.
[b] Average: $T_g = W_1 T_{g1} + W_2 T_{g2}$.
[c] $[T_g - (T_g)_{av}]/(T_g)_{av} = -\theta/(1 - \theta)$.

$X'_c = -\theta$, for $\theta \ll 1$. The quantities become functionally identical if F_x goes to zero.)

The experimental values of T_g (expt) lie in between the T'_gs estimated from equations (6.41) and (6.42), so that, apparently, neither the increase in physical crosslinking effect, nor the decrease in hydrogen bonding effect plays a decisive role.

Noting the distinct possibility of extensive grafting in the homogeneous materials shown in Table 6.6,[61] Klempner *et al.*[64] repeated portions of that work using an acrylic especially prepared to reduce the role of grafting. The results are shown in Table 6.7. Two glass transitions are observed, and the material is heterophase in nature. However, the T_g's are shifted inwards, indicating some molecular mixing of the two polymers. The authors[64] remark that, depending on the thermal history, sometimes an intermediate

Table 6.7. Glass Transition Temperatures of PU/PA SINs, °K[64]

Composition, percent						
PU	100%	80	60	40	20	0
PA	0	20	40	60	80	100
State of crosslinking						
Full SIN	—	317.2	318.5	319	319	320.5
	208	209	211.2	212	212.5	—
PDIPN-1[a]	—	313	314	316	318	318
	208	209	210	211.5	212.5	—
PDIPN-2[b]	—	318	318	319	319.5	320.5
	210	210.5	211	211	211.5	—
Chemical blend	—	306	306	307.5	312	318
	210	209	209	210	211.5	—

[a] Crosslinked PU, linear PA.
[b] Crosslinked PA, linear PU.

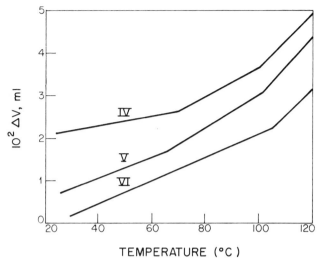

Figure 6.20. Dilatomeric data for poly(vinyl trichloroacetate)/poly(methyl methacrylate) AB crosslinked copolymers. Networks IV and V show two second-order transitions, while network VI shows only one transition.[65]

T_g was observed for the full IPNs, indicating the presence of a possible third phase.

Traditionally, the classical method of observing transitions has been through volume-temperature studies. Bamford et al.[65] employed the dilatometric technique for the study of glass transitions in ABCPs (see Figure 6.20). Changes in slope* indicate a second-order transition, in this case a glass–rubber transition. While in principle the complete set of transitions can be observed, low sensitivity combined with lack of valuable secondary information has apparently caused this method to fall into disfavor among polymer chemists. However, Figure 6.20 clearly shows the shift in T_g with composition, and the slope change indicates the relative intensity of the transition.

Recently, Lipatov et al.[66,67] combined broad-line NMR spectroscopy and inverse gas chromatography to study the intermediate glass transition and the activation energies of the T_g's. They prepared a sequential IPN using a polyurethane as polymer I and polystyrene as polymer II. Lipatov et al. first measured the NMR signal as a function of temperature (see Figure 6.21)[67] The spectrum shape changes from a simple one-component line to a more complicated two-component line as the temperature is increased,

---

* The change in slope in a volume–temperature plot actually means a change in the volume coefficient of expansion, and usually indicates a second-order transition of some type. The glass–rubber transition is a special type of second-order transition. A first-order transition (melting, boiling) has a discontinuity in the V–T plot.

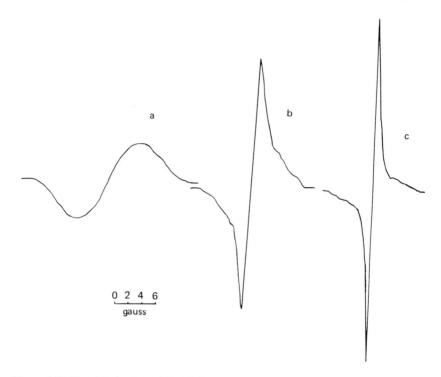

Figure 6.21. Signal derivatives of the NMR spectrum for a polyurethane/polystyrene IPN at $w_2/w_1 = 0.174$ at (a) $-150°C$; (b) $-30°C$; (c) $+25°C$.[67]

indicating regions of various mobilities. The second moments, ΔH_2^2, and line widths δH were determined and plotted against temperature. The quantity δH stayed broader for the IPNs up to 90°C, indicating altered mobility.

The quantity ΔH_2^2 is plotted vs. temperature in Figure 6.22[67] For the IPN plotted in Figure 6.22, as well as the others investigated, three glass transition temperatures were noted. The transitions near $-70°C$ and $+60°C$ coincide with transition points of the individual networks as independent components of the system. However, the appearance of a third, inter-mediate transition in the range $+25-60°C$ indicates the presence of an interphase region in these materials. It should be noted that intermediate transitions have already been discovered in block copolymers.[68]

It is known that for two-component systems, additivity of lines takes place in the NMR spectra. Assuming that the same additivity should be observed for the second moments, ΔH_2^2 may be represented by

$$[\Delta H_2^2]_{1,2} = W_1[\Delta H_2^2]_1 + W_2[\Delta H_2^2]_2 + [\Delta H_2^2]_i \qquad (6.50)$$

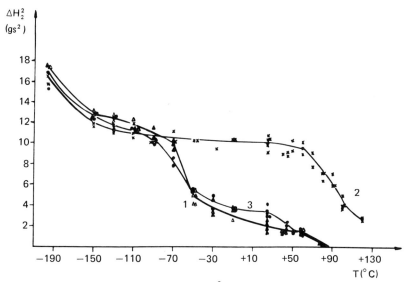

Figure 6.22. Temperature dependence of $\Delta H_2{}^2$ for (1) PU; (2) styrene–divinylbenzene copolymer; (3) IPN at $w_2/w_1 = 0.174$. [67]

where the W's represent weight fractions, and $[\Delta H_2^2]_i$ is a term responsible for interaction, if any. When interactions are absent on a molecular level, $[\Delta H_2^2]_i$ equals zero, which was found experimentally. Presumably, this indicates an absence of chemical interactions, since the presence of an intermediate, third phase indicates a definite physical interaction.

Inverse gas chromatography depends on putting a vapor on the surface or interior of the sample and measuring the volume retained as a function of temperature, sometimes called the retention volume, V_g.[69] A somewhat idealized result is shown in Figure 6.23. The enthalpy of adsorption or solution, ΔH, can be obtained from the straight line portions from

$$\frac{\partial \ln V_g}{\partial(1/T)} = -\frac{\Delta H}{R} \tag{6.51}$$

Lipatov *et al.*[64] calculated the excess enthalpy of mixing at the point of equilibrium absorption according to the equation

$$\Delta H = R \frac{\partial[-\ln V_g P_1^0 - (P_1^0/RT)(B_{11} - V_1)]}{\partial(1/T)} \tag{6.52}$$

where P_1^0 represents the saturated vapor pressure of the sorbate, R the gas constant, B_{11} the second virial coefficient, and V_1 the molar volume of the sorbate at the temperature of the column.

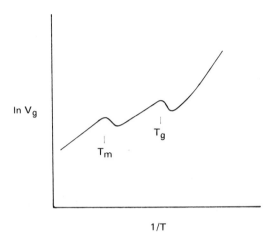

Figure 6.23. An idealized retention volume plot, showing a glass transition and a melting transition.

The experimental results of the retention volume experiment are shown in Figure 6.24.[66] Three transition temperatures may be seen. The intermediate transition of interest occurs in the range of 44–78°C, between the transition temperatures of the two independent networks. This transition is explained by Lipatov et al.[67] as a true interpenetrating or intermediate region at the interface between the two networks, the basis for a third phase. Again, such interphase compositions come to the fore because of the large size of the molecules, which approaches the dimensions of the phases themselves.

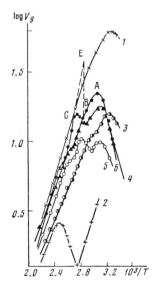

Figure 6.24. Dependence of log V_g on $1/T$ for, 1, the first network and, 2, the second network in polyurethane/polystyrene sequential IPNs containing: 3, 0.09; 4, 0.384; 5, 0.483; and 6, 0.567 fraction by weight of the second network. Sorbate for sample 1, benzene; for the remainder, acetone.[66]

The enthalpy of mixing of the networks, calculated from equation (6.52), is shown in Figure 6.25.[66] At low concentrations of network II, ΔH_{12} is observed to be negative. Although it is not a general criterion, thermodynamic mutual solution may be possible in this region. However, the enthalpy of mixing increases as the concentration of the second network is increased and becomes positive at a value of $W_2 = 0.3$.

Also shown in Figure 6.25 is the proportion of material in the transition region, q, which was estimated from

$$q = \left[\frac{V_g(E) - V_g(B)}{V_g(E)} \right] \times 100 \qquad (6.53)$$

where $V_g(E)$ is the total value of the sorbate volume retained by the molten first network and the intermediate region, and $V_g(B)$ is the volume retained only by the first network. Note points E and B at the top and bottom of the transition, respectively, in Figure 6.24. The increase in ΔH above zero in Figure 6.25 is accompanied simultaneously by an increase in q, that is, by an increase in the boundary region where thermodynamic incompatibility is observed. Thus, the heat of mixing, a strictly thermodynamic quantity, indicates the emergence of phase boundary material.

It is observed that Figure 6.25 only delineates the lower half of the concentration region, taking polymer II from zero to midrange concentration. Thus, the quantity of interfacial boundary material may be increasing partially at the expense of the absolute amount of polymer II present, as well as changes in ΔH_{12}.

Lipatov et al.[65] also calculated the activation energies, E_a, of the IPN and homopolymer transitions, as shown in Table 6.8. Analysis of the data indicates that the introduction of the second network (PS) into the first (PU) leads to a decrease in E_a in comparison with the individual networks.

Lipatov et al.[67] also postulated that the intermediate region has a loosely packed structure, leading to a shift to lower temperatures at the

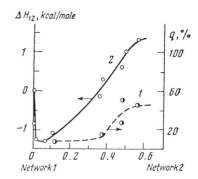

Figure 6.25. Polyurethane/polystyrene sequential IPNs. Dependence of: 1, the proportion of the intermediate layer, q, and 2, the enthalpy of mixing of the networks, on the concentration of the second network in IPN.[66]

Table 6.8. Activation Energies of Temperature Transitions in Poly-
urethane/Polystyrene IPNs and Homopolymers[67]

	E_a, kcal/mole		
	Transition $> -70°C$	Transition $> +25°C$	Transition $> +60°C$
Network I	12.5	—	37.2[a]
IPN $w_2/w_1 = 0.119$	5.8	21.4	25.2
$w_2/w_1 = 0.174$	5.3	16.7	21.8
$w_2/w_1 = 0.242$	4.2	15.7	23.2
Network II	—	—	12.3

[a] Caused by crystallinity in the PU.

beginning of each transition. This result is particularly interesting in the light of the work by Helfand and co-workers[70-73] which predicts a rarefication of mass at the interface of incompatible blends. This rarefication is caused, fundamentally, by the positive heat of mixing and loss of conformational entropy within the region of intermolecular contact.

As attention is turned to the results of evaluating T_g by mechanical methods, the postulated existence of the interphase component will be given further examination.

6.7.2.3. Results of Mechanical Methods

A principal value of evaluating glass–rubber transitions via mechanical methods stems from the fact that a direct readout of the modulus or hardness of the material vs. temperature is obtained simultaneously. Since many instruments also yield the loss modulus or similar parameters, the results are useful for evaluating engineering properties of the materials.

Several instruments are available. The Gehman[74] and Clash–Berg[75] instruments yield only the shear modulus, G. Particular advantages include low cost, speed, and simplicity. The torsional pendulum[76] and torsional braid[77] yield both storage and loss parameters, but the torsional braid is sometimes difficult to use in conjunction with IPNs since the polymer must be dipped onto the braid in the linear form or polymerized *in situ*. A widely employed instrument is the Rheovibron,* invented by Takayanagi, which elongates a strip of material in a cyclic manner, yielding the complex modulus, E^*, and the loss tangent, tan δ. The loss modulus, $E''=E^*\sin \delta$, and the storage modulus, $E' = E^* \cos \delta$, can be calculated.

* Toyo Measuring Instruments Co. Ltd., Tokyo.

These latter experiments are sometimes referred to as dynamic mechanical spectroscopy, DMS.

Using the Rheovibron, Huelck *et al.*[17] measured E' and E'' as a function of temperature for a series of sequential IPNs, see Figure 6.26. Since the T_g's of polystyrene and poly(methyl methacrylate) are nearly equal (100°C and 107°C at 0.1 Hz, respectively), Figures 6.26a and 6.26b provide an opportunity for comparing the effect of compatibility on the behavior of the glass transition, with minimal disturbance to the other properties of the system. It should be noted, as expressed in Section 5.2.1, that poly(ethyl acrylate) and poly(methyl methacrylate) are chemically isomeric. Figure 6.26a illustrates a typical incompatible pair, showing two distinct glass transitions with some inward shifting. On the other hand, Figure 6.26b shows only one broad transition, typical of semicompatible systems.

Figure 6.26 bears comparison to Figure 6.2. The fine structure in Figure 6.2a is now to cause a smearing of the two glass transitions, until they appear to merge into one extraordinarily broad transition, Figure 6.26b. The original interpretation of this single broad transition was that different regions of space had different compositions, each yielding its own glass transition temperature.[78] The original Rouse–Bueche theory[79,80] requires approximately 50 mers to undergo coordinated motion for the glass transition relaxation phenomenon to occur. This corresponds to a volume of

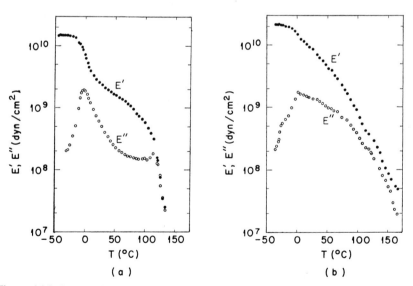

Figure 6.26. Storage (E') and loss (E'') moduli: (a) 48.8/51.2 PEA/PS; (b) 47.1/52/9 PEA/PMMA IPNs, both at 110 Hz.[17]

about 10,000 Å^3. Because of the chain characteristics of polymers, however, volumes of 100,000 Å^3 or more will be required to average out random concentration fluctuations for even ideally compatible IPNs. If the minimum volume required for independent contributions to the relaxation spectrum is the same or smaller than that required to yield homogeneous overall compositions, a broadened transition will result.

Alternately, Figure 6.26b bears interpretation along the reasoning of Lipatov et al.,[66,67] where a centrally located T_g arises from extensive interfacial boundary material. Since local concentration fluctuations would be expected to be rife in the phase boundary region (or interphase) at a level of 50–100 Å, the whole material may be considered a macroscopic interphase structure, and the two models portray different concepts of the same phenomenon in this case.

A model for Figure 6.2a consistent with Figure 6.26b is shown in Figure 6.27. The illustrated network II domains form the less continuous phase. Network I, of course, occupies the remaining space.

In one of the few studies that employed more than one instrument to study the T_g's of IPNs and related materials, Kim et al.[55] combined DSC and torsional pendulum to study a series of SINs. In Table 6.9, polyurethane is the low T_g network and poly(methyl methacrylate) or polystyrene is the high T_g network. Semi-SINs (Kim et al. prefer the term "pseudo" instead of "semi") and chemical blends are also shown.

A comparison of experimental techniques is essential for understanding the slight differences in the results. The maximum in G'' at 0.2 Hz was taken as T_g in the torsional pendulum study. The DSC scanning rates were 10°C/min. Since a factor of 10 in the frequency raises T_g about 6–7°C, if the torsional pendulum frequency had been increased from 0.2 Hz to about 1.0 Hz, the values in Table 6.9 would be predicted to match virtually identically. (Raising the temperature at 1°C/min, in volume–temperature studies often matches shear modulus studies using 10 sec or 0.1 Hz as the deformation time.)

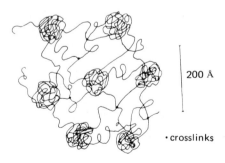

200 Å

• crosslinks

Figure 6.27. Model of the interpenetration mode through the cell-wall structures. Note that the fine structures have dimensions smaller than the end-to-end distances of most macromolecules.[17]

Table 6.9.[55] Glass Transition Temperature, T_g (°K), for SIN

| Composition | Low T_g | | High T_g |
	DSC	Torsion	DSC
Homopolymers			
UC100	224	219	
UL100	224	218	
MC100			382
ML100			382
SC100			375
SINs			
UC85MC15	228	229	
UC75MC25	230	227	
UC60MC40	226	225	366
UC40MC60	227	227	367
UC25MC75	224		371
UC15MC85	224		379
UC80SC20	228		372
UC75SC25	226		372
UC60SC40	226		366
UC20SC80	228		369
Semi-SINs			
UC75ML25	226	219	
UC75MC25	224	219	
Chemical blends			
UL75ML25	224	219	

In general, the T_g's in Table 6.9 are shifted inward. The shift is more prominent in the high T_g component than in the low T_g component, and more prominent in the polyurethane/poly(methyl methacrylate) pair than in the polyurethane/polystyrene pair. Kim et al.[55] note that the solubility parameter of poly(methyl methacrylate) is closer to polyurethane than that of polystyrene. The relative changes in T_g are consistent with the observed morphologies (see Figure 6.4).

A comparison of the UCSC density changes shown in Figure 6.18 with the decreasing T_g values noted in Table 6.9 leads to a seeming contradiction. A density increase, of course, implies a decrease in free volume, and hence an increase in T_g. The apparent dilemma may be resolved by two nonincon-sistent considerations: (1) the actual molecular interpenetration may lead to a dependence of T_g based on equation (6.41), and/or (2) lacking a densification, the polystyrene T_g would have decreased even further.

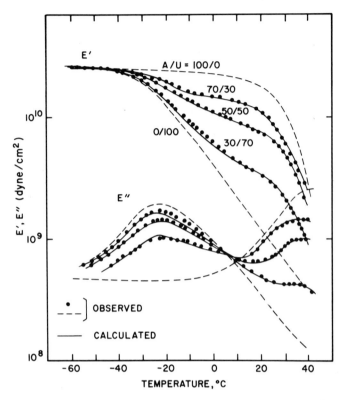

Figure 6.28. Temperature dependences of the dynamic storage modulus E' and the loss modulus E'' of the interpenetrating elastomeric networks: (●) observed values; (- - -) component homopolymers.[36]

The effect of composition on the modulus for acrylic–urethane IENs is shown in Figure 6.28.[36] Two distinct transitions are observed, with the acrylic copolymer softening at the higher temperature. The slight shifting and broadening of the glass transitions indicate a modest degree of molecular mixing. However, the large shifts in E' indicate changes in phase continuity. The reader should compare the moduli in Figure 6.28 with the morphology of IENs shown in Figure 6.7.

Values of storage modulus at 23°C are plotted as a function of composition in Figure 6.29. Takayanagi's parallel model for the mechanical behavior of a two-component system[82] corresponds to the case in which the stiffer component is continuous, while his series model corresponds to the case in which the softer component is continuous. Clearly the experimental results agree best with the parallel model over most of the concentration

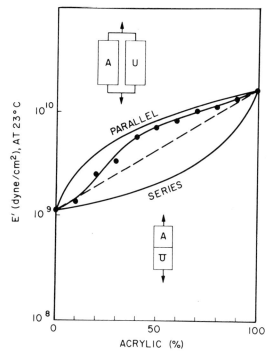

Figure 6.29. Plot of the dynamic modulus E' at 23°C against the polyacrylate content. The upper and the lower solid lines are calculated with a parallel and a series model, respectively.[36]

range. The observations (based on Figure 6.7) confirm that a phase inversion takes place at about 30% acrylic component, however.

The behavior of castor oil–urethane/polystyrene sequential IPNs was investigated by Yenwo et al.[47] In Figure 6.13 these materials were shown to have small phase domains of 300–500 Å, caused by the rather high crosslink level of the castor oil–urethane elastomer used as polymer I. The extent of mixing is also illustrated by its dynamic mechanical behavior, which is summarized in Table 6.10.[47] Like much of the previous data, a clear inward shift of the T_g's is noted. In the following, the extent of molecular mixing implied by Table 6.10 will be put on a quantitative basis.

6.7.2.4. Mass Balance Analysis

In Tables 6.9, 6.10, and elsewhere, a definite inward shifting of the T_g's was noted. This inward shift was qualitatively interpreted in terms of partial

Table 6.10. Castor Oil-Urethane/PS (CO–U/PS) IPN Glass Transition
Temperatures[47]

Sample	NCO/OH	Cross linker	IPN Composition CO–U/PS	T_g, °C CO–U	PS
1	—	1% DVB	0/100	—	104
2	0.75	2,4 TDI	100/0	−24	—
3	0.75		30/70	−4	100
4	0.75	1% DVB	44/56	0	96
5	0.85	2,4 TDI	100/0	−12	—
6	0.85		34/66	−8	92
7	0·85	1% DVB	50/50	8	88
8	0.95	2,4 TDI	100/0	−4	—
9	0.95		37/63	60	60
10	0.95		44/56	42	42
11	0.75	80/20:2,4/2,6 TDI	100/0	−20	—
12	0·75		31/69	−12	92
13	0·75		51/49	0	88
14	0.85	80/20:2,4/2,6 TDI	100/0	− 6	—
15	0.85		38/62	28	80
16	0.85		46/54	60	60
17	0.95	80/20:2,4/2,6 TDI	100/0	0	—
18	0.95		40/60	32	76
19	0.95		46/54	48	48
20	0.75	HDI	100/0	−42	—
21	0.75		29/71	−20	100
22	0.75		31/69	−18	98
23	0.85	HDI	100/0	−38	—
24	0.85		33/67	−8	80
25	0.85		35/65	0	78
26	0.95	HDI	100/0	−26	—
27	0.95		36/64	4	72
28	0.98		38/62	8	70

mutual solution of the two polymers. If it is assumed that the random copolymer equations (6.41) and (6.42) hold, then the actual composition within each phase may be calculated. Of course, it must also be assumed that the composition everywhere within each phase is constant, otherwise some type of average value will be obtained.

The above results can be combined with electron microscopy studies to yield a complete mass balance of the material. There is a theorem in composite systems which states that the cross-sectional area of the dispersed phase is proportional to its volume.[82] Through use of this theorem, the mass fraction of each phase can be determined by weighing cut out portions of electron micrographs. From a knowledge of both that and the phase composition, a complete mass balance analysis can be carried out.

Such an experiment was done by Scarito and Sperling,[83] using an epoxy/poly(*n*-butyl acrylate) SIN, grafted together with variable levels of glycidyl methacrylate. The scanning electron micrographs shown in Figure 6.30 were obtained using an etching technique. Figure 6.30 provided a quantitative basis for evaluating the fraction of each phase (see Table 6.11). Since the overall composition was 80/20 epoxy/acrylic in each case, the effects of mixing are immediately obvious in Table 6.11. From the inward shift of the T_g's afforded through DMS studies, the composition within each phase was estimated (see Table 6.12). Also shown in Table 6.12 is a complete mass balance calculated from the fraction of each polymer within each phase. Ideally, of course, these ought to come out to the known 80/20 composition; the differences form a measure of the experimental and theoretical errors. Thus, not only are morphology and glass transition studies qualitative guides to the molecular mixing, but they can be used quantitatively as well.

6.8. TRANSITION LAYER THICKNESS

In a recent study, Shilov *et al.*[84] employed a small–angle x-ray scattering SAXS technique to characterize the SIN microheterogeneities. A polyurethane/polyurethane–acrylate SIN composition was formed by mutual solution of the two prepolymers into a monolithic film, followed by simultaneous curing via independent routes.

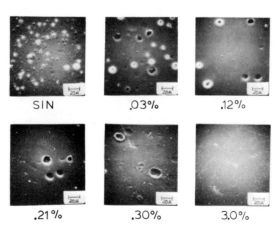

Figure 6.30. Effect of glycidyl methacrylate grafting on the morphology of epoxy/PnBA SINs.[83]

Table 6.11. Phase Volume Fraction of Epoxy/Acrylic
SIN[83]

% G.M.[a]	Fraction Dispersed Phase	Fraction Matrix Phase
0.0	0.16	0.84
0.0	0.18	0.82
0.03	0.11	0.89
0.03	0.12	0.88
0.12	0.04	0.96
0.21	0.03	0.97
0.30	0.04	0.96
0.30	0.03	0.97
3.0	0	1.00
3.0	0	1.00

[a] Glycidyl methacrylate, grafting agent.

The basis of the experiment involves the mean square of the electron density fluctuation, $\langle \Delta \eta^2 \rangle$. The scattering intensity arising is then extrapolated to zero angle, permitting determination of the thickness of the transition layer between the two phases, E, the distance of heterogeneity, l_c, and the average radii of the largest heterogeneous regions, R.[85-87]

Values of l_c and R are shown in Table 6.13.[84] The values average between a few hundred and several hundred angstroms, with a maximum at midrange compositions, corroborating the results of electron microscope studies shown on other materials.

The method of Vonk,[87] which depends on the analysis of the tail of the SAXS pattern, was used to calculate the thickness of the transition layer. Values of 20–40 Å were obtained. These thicknesses are comparable to interphase thicknesses estimated for block copolymers (see Section 2.2). Thus, an actual numerical estimate of the extent of interpenetration has been achieved. Ultimately, such results might provide better insight into the differences illustrated in Figures 6.26a and 6.26b.

6.9. PHASE CONTINUITY EVALUATION VIA MODULUS

Modulus–temperature studies yield information about phase continuity qualitatively. Idealizing slightly, Figures 6.26 and 6.28 show a dip in E' at the first T_g, followed by a plateau, followed by a second dip at the second T_g. High values of the intermediate plateau are usually associated with greater continuity of the higher T_g component.

Table 6.12. Phase Composition of Epoxy/Acrylic SIN[(83)]

| % G.M.[a] | T_g °C | | Dispersed phase, wt. frac. | | | Matrix phase, wt. frac. | | | Calc. PnBA,[b] total | Calc. epoxy,[c] total |
	Acrylic	Epoxy	Equation	PnBA	Epoxy	PnBA	Epoxy			
0.0	−59	82	6.41	0.97	0.03	0.09	0.91		0.23	0.77
			6.42	0.98	0.02	0.15	0.86		0.28	0.72
0.30	−42	65	6.41	0.82	0.18	0.12	0.88		0.15	0.86
			6.42	0.89	0.11	0.19	0.81		0.22	0.78
3.0	—	56	6.41	—	—	0.30	0.70		0.30	0.70
			6.42	—	—	0.20	0.80		0.20	0.80

[a] Glycidyl methacrylate, grafting agent.
[b] Actual fraction = 0.20.
[c] Actual fraction = 0.80.

Table 6.13. Parameters of Heterogeneity in Polyurethane/Polyurethane-Acrylate SIN[84]

Network composition	l_c, Å average distance of the heterogeneity	$\langle \Delta \eta^2 \rangle$, e^2 mol^2/cm^6 difference of electron density	R, Å largest average radii of heterogeneous regions
PU	550	1.273×10^{-6}	
IPN			
PU:PUA 95:5	440	451.8×10^{-6}	360
IPN PU:PUA 90:10	475	507.6×10^{-6}	480
IPN PU:PUA 60:40	454	585.2×10^{-6}	450
IPN PU:PUA 50:50	533	602.04×10^{-6}	730
IPN PU:PUA 25:75	123	$1284\,13 \times 10^{-6}$	430
IPN PU:PUA 10:90	134	967.1×10^{-6}	540
PUA	280	1.095×10^{-6}	

More quantitative information depends on the use of models. The Takayanagi models were already mentioned in connection with Figure 6.29. More analytical models have been evolved by Kerner,[88] Hashin and Shtrikman,[89] and Davies.[90] Briefly, the first two theories[88,89] assume spherical particles dispersed in an isotropic matrix. From the modulus of each phase, the composite modulus is calculated. An upper or lower bound modulus is arrived at by assuming the higher or lower modulus phase to be the matrix, respectively. The theory is reviewed elsewhere.[1,3]

Dual-phase continuity was assumed by Davies.[90] The shear modulus for a Davies composite may be expressed

$$G^{1/5} = v_1 G_1^{1/5} + v_2 G_2^{1/5} \tag{6.54}$$

where v is the volume fraction of phase 1 or 2 as indicated.

Dickie et al.[91] examined the composite behavior of semi-I and semi-II latex IPNs, as well as their physical blends. Their fit to the Kerner equation is shown in Figure 6.31. The HLP-1 (glassy component as polymer II, see section 5.4.3.2) materials follow the Kerner upper bound model, while the HLP-2 materials more nearly follow the lower bound model. This indicates that in latex IPNs, the second polymerized component is more continuous in space, after molding, than the first polymerized component. This is consistent with the core–shell model, see Figure 6.12.

Allen, et al.[92] evaluated their polyurethane/poly(methyl methacrylate) semi-SINs (interstitial composites) in terms of the Davies, Kerner, and Hashin and Shtrikman theories (see Figure 6.32). The Davis theory, equation (6.54), clearly fit best, indicating that the material most probably has dual phase continuity. The actual morphology of one of these compositions was shown in Figure 6.3.

Donatelli *et al.*[93] also applied the Kerner and Davies theories to their sequential IPNs and found that the Davies equation fits best (see Figure 6.33). Two of these compositions, shown in the lower part of Figure 2.3, suggest dual-phase continuity in a qualitative way. Semi-II IPNs, illustrated in the center of Figure 2.3, fell below the Davies line, suggesting one continuous and one discontinuous phase. While the Dickie *et al.* latex materials would not be expected to exhibit dual-phase continuity, it is interesting that both the Allen *et al.* and the Donatelli *et al.* materials do.

An interesting dual-phase continuity model for the modulus of an IPN has recently been suggested by Paul,[94] Figure 6.34. As shown in the insert, the model consists of a regular lattice of cubes, each containing a regular arrangement of the two phases. The model is based on the work of Kraus and Rollman[95] who proposed an extension of the Takayanagi models.[81] Via the Takayanagi approach, the approximate upper and lower bounds for the IPN modulus were estimated, as illustrated by the shaded zone in Figure 6.34. For comparison, Paul also shows the values obtained on parallel, series, and dispersion models.

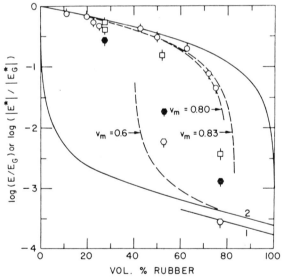

Figure 6.31. Poly(methyl methacrylate)/poly(butylacrylate) latex semi-IPN. Dependence of tensile modulus of compression-molded specimens on latex particle composition and type: E_g is Young's modulus; open squares, physical blends; open circles, HLP1, open hexagons, HLP2 with BDMA in first stage; filled hexagons, HLP2 with BDMA in second stage; curve 1, based on BA homopolymer data; curve 2, based on BA/BDMA copolymer data; broken curves, calculated with V_m as noted.[91]

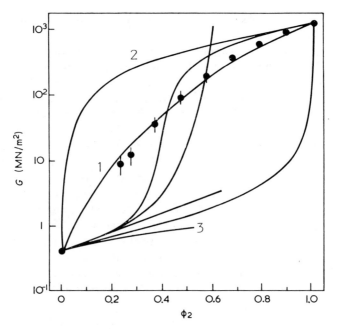

Figure 6.32. Log$_{10}$ (shear modulus) G as a function of volume fraction (ϕ_2) of PMMA for PMMA–PU interstitial composites. Filled circles are experimental points, and numbers on theoretical curves refer as follows: (1) Davies[90]; (2) Kerner[88]; (3) Hashin and Shtrikman[89]; other lines refer to earlier theories.[92]

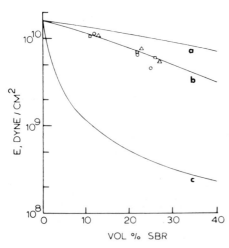

Figure 6.33. Modulus–composition curves for SBR/PS semi-I and IPN materials.[93] (a) Kerner equation (upper bound); (b) Davies equation; (c) Kerner equation (lower bound). (○) Series 3 (semi-1); (▽) Series 4 (full); (□) Series 5 (full).

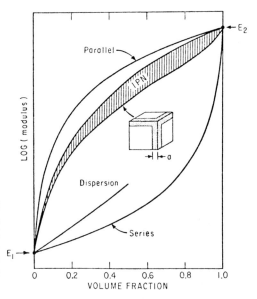

Figure 6.34. Mechanical modulus of various polymer blend phase arrangements. The cube shows a unit cell model of an idealized IPN structure. The shaded area is determined by upper- and lower-bound estimates for this model.[94]

The placement of the shaded zone in Figure 6.34 predicts a modulus for IPNs only slightly below the parallel arrangement; i.e., that IPNs exhibiting two continuous phases ought to be significantly stiffer than otherwise expected. It is interesting to note that the results of Huelck *et al.* (see Figure 6.26, for example)[17] fit the IPN region in Figure 6.34 almost exactly.

An especially interesting model that predicts dual-phase continuity is based on work by Budiansky.[96] For two-phased systems, the Budiansky equation may be expressed

$$\frac{v_1}{1 + \varepsilon(G_1/G - 1)} + \frac{v_2}{1 + \varepsilon(G_2/G - 1)} = 1 \qquad (6.55)$$

where

$$\varepsilon = \frac{2(4 - 5\nu_\rho)}{15(1 - \nu_\rho)} \qquad (6.56)$$

The quantity ν_ρ represents Poisson's ratio for the composite having a shear modulus G, and G_1 and G_2 are the shear moduli of the component polymers having volume fractions v_1 and v_2, respectively.

It is interesting to note that equation (6.55) is symmetrical, and the reversal of the geometric roles of components 1 and 2 does not change the composite property, provided its volume fraction remains the same. The Budiansky equation assumes cocontinuity of both polymers, in a manner

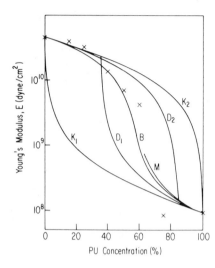

Figure 6.35. Young's modulus vs. polyurethane concentration for the polyurethane–poly(methyl methacrylate) SINs at 23°C. Solid lines are based on the theoretical models with K_1, the Kerner equation, assuming the polyurethane as the continuous phase; D_1 and D_2 the respective Dickie equations, M the Mooney equation; and B, the Budiansky equation.[97]

similar to the Davies equation [equation (6.54)], but may be more sensitive than the latter because the modulus is not reduced to the 1/5th power.

In a recent study of the viscoelastic behavior of polyurethane/poly (methyl methacrylate) SINs, Kim *et al.* (97) evaluated the several equations treated above. They found that the Budiansky equation fit their data best (see Figure 6.35). The S-shape character of the Budiansky curve suggests a phase inversion, which was experimentally established by these workers (see Figure 6.4). For sake of comparison, the Kerner, Mooney, and Dickie equations are also illustrated.

6.10. THE INTERPENETRATION PROBLEM

Because electron micrographs and glass transition studies as such cannot prove interpenetration, some modeling of the data is necessary to arrive at a coherent picture. While some of the quantitative aspects of the interpenetration and phase continuity were discussed in Chapter 4 on homo-IPNs, a qualitative review of the evidence is nevertheless instructive.

(a) Certainly the two-phased nature of IPNs indicates that interpenetration exists on a supermolecular scale, rather than on a true molecular scale. Evidence for the existence of trapped or dissolved molecular segments, other forms of interpenetration, is given by the unexpected inward shifting of the two glass transitions.

(b) The first network in a sequential IPN is clearly continuous on a macroscopic scale. This is obvious from both a simple inspection of the sample during IPN synthesis and electron microscopy observations that the first network always forms the matrix.

(c) Although the bulk of the second network in incompatible IPNs appears to lie within the cells, the 100-Å fine structure within the matrix is a clue to the continuity of the second network. A proposed model of the fine structure pervading the cell walls and connecting the cell contents is shown in Figure 6.27.

(d) The fine structure exhibits domain dimensions of the order of 100 Å or less. This is the same order of magnitude as the expected end-to-end distances of a polymer segment between crosslink junctures (ca. 200 mers) and much smaller than the end-to-end distance of the primary chains. If a model similar to Figure 6.27 is accepted, the phases need not be continuous, but the components must be if the second polymer formed a reasonable network. It should be pointed out that the domains observed here are much smaller than the interpenetrating elastomeric networks of Frisch (Figure 6.7). The phase domain size of sequential IPNs is controlled primarily by the crosslink density and compatibility of the two polymers. The corresponding phase domain size of the IENs is controlled by the initial latex particle dimensions.

(e) The evidence presented in Figures 6.32 and 6.33 appears consistent with the notion of two cocontinuous phases. While many of the domains appear spherical there is growing evidence that many of them may be cylinders or even connected cylindrical structures.

(f) The presence of crosslinking, especially in network I, controls and reduces the phase domain size. Thus, block and graft sites have a counterpart in the IPN dual crosslinked structure.

(g) Chains entrapped physically by network forces (interpenetrating macrocatenated structures) influence the free energy of mixing, although the quantitative role is not yet well understood.

(h) Lastly, the relationship between IPN crosslinking and polymer blend mutual solubility needs discussing. There is some evidence that IPN formation may reduce mutual solubility. Two cases may be considered. (a) Chains may be trapped in separate regions of space (7), or (b) because of the apparent increase in molecular weight on crosslinking, the entropic component of the free energy of mixing decreases. This last may be responsible, in part, for the fine structure in the homo-IPNs (see Figure 4.5).

REFERENCES

1. C. B. Bucknall, *Toughened Plastics*, Applied Science Publishers, London (1977).
2. A. Noshay and J. E. McGrath, *Block Copolymers—Overview and Critical Survey*, Academic Press, New York (1977).

3. J. A. Manson and L. H. Sperling, *Polymer Blends and Composites*, Plenum, New York (1976).
4. Yu. S. Lipatov and L. M. Sergeeva, *Interpenetrating Polymeric Networks*, Naukova Dumka, Kiev (1979).
5. D. Klempner, *Angew. Chem.* **90**, 104 (1978).
6. L. H. Sperling, *J. Polym. Sci. Macromol. Revs.* **12**, 141 (1977).
7. H. L. Frisch, K. C. Frisch, and D. Klempner, *Mod. Plast.* **54**, 76, 84 (1977).
8. D. R. Paul and S. Newman, *Polymer Blends*, Vols. I and II, Academic, New York (1978).
9. L. H. Sperling, in *Polymer Blends*, Vol. 2, D. R. Paul and S. Newman, eds., Academic, New York (1978).
10. K. Kato, *Jpn Plast.* **2** (April), 6 (1968).
11. M. Matsuo, *Jpn Plast.* **2** (July), 6 (1968).
12. D. A. Thomas, *J. Polym. Sci.* **60C**, 189 (1977).
13. S. C. Kim, D. Klempner, K. C. Frisch, H. L. Frisch, and H. Ghiradella, *Polym. Eng. Sci.* **15**, 339 (1975).
14. J. Brandrup and E. H. Immergut, eds., *Polymer Handbook*, 2nd ed., Wiley-Interscience, New York (1975).
15. B. A. Ozerkovskii, Yu. B. Kalymkov, U. G. Gafurov, and V. P. Roschchupkin, *Vysokomol. soyed.* **A19**(7), 1437 (1977).
16. H. Keskkula and P. A. Traylor, *J. Appl. Polym. Sci.* **11**, 2361 (1967).
17. V. Huelck, D. A. Thomas, and L. H. Sperling, *Macromolecules* **5**, 340, 348 (1972).
18. A. A. Donatelli, D. A. Thomas, and L. H. Sperling, in *Recent Advances in Polymer Blends, Grafts, and Blocks*, L. H. Sperling, ed., Plenum, New York (1974).
19. S. Krause, *Macromolecules* **3**, 84 (1970).
20. D. J. Meier, *J. Polym. Sci.* **26C**, 81 (1969).
21. A. A. Donatelli, L. H. Sperling, and D. A. Thomas, *J. Appl. Polym. Sci.* **21**, 1189 (1977).
22. G. Allen, M. J. Bowden, D. J. Blundell, F. G. Hutchinson, G. M. Jeffs, and V. Vyvoda, *Polymer* **14**, 597 (1973).
23. G. Allen, M. J. Bowden, O. J. Blundell, G. M. Jeffs, J. Vyvoda, and T. White, *Polymer* **14**, 604 (1973).
24. C. H. Bamford, G. C. Eastmond, and D. Whittle, *Polymer* **16**, 377 (1975).
25. G. C. Eastmond and E. G. Smith, *Polymer* **17**, 367 (1976).
26. D. J. Meier, *Polym. Prepr.* **11**, 400 (1970).
27. S. C. Kim, D. Klempner, K. C. Frisch, W. Radigan, and H. L. Frisch, *Macromolecules* **9**, 258 (1976).
28. N. Devia-Manjarres, J. A. Manson, L. H. Sperling, and A. Conde, *Polym. Eng. Sci.* **18**, 200 (1978).
29. N. Devia, J. A. Manson, L. H. Sperling, and A. Conde, *Macromolecules* **12**, 360 (1979).
30. N. Devia, J. A. Manson, L. H. Sperling, and A. Conde, *Polym. Eng. Sci.* **19**(12), 869 (1979).
31. N. Devia, J. A. Manson, L. H. Sperling, and A. Conde, *Polym. Eng. Sci.* **19**(12), 878 (1979).
32. N. Devia, J. A. Manson, L. H. Sperling, and A. Conde, *J. Appl. Polym. Sci.* **24**, 569 (1979).
33. C. B. Bucknall and T. Yoshii, *Br. Polym. J.* **10**, 53 (1978).
34. R. E. Touhsaent, D. A. Thomas, and L. H. Sperling, *J. Polym. Sci.* **46C**, 175 (1974).
35. R. E. Touhsaent, D. A. Thomas, and L. H. Sperling, in *Toughness and Brittleness of Plastics*, R. D. Deanin and A. M. Crugnola, eds., Advances in Chemistry Series No. 154, American Chemical Society, Washington, D.C. (1976).
36. M. Matsuo, T. K. Kwei, D. Klempner, and H. L. Frisch, *Polym. Eng. Sci.* **10**, 327 (1970).

37. D. Klempner, H. L. Frisch, and K. C. Frisch, *J. Polym. Sci. A-2* **8**, 921 (1970).

38. K. Kato, *Polym. Lett.* **4**, 35 (1966).

39. J. Sionakidis, L. H. Sperling, and D. A. Thomas, *J. Appl. Polym. Sci.* **24**, 1179 (1979).

40. L. H. Sperling, Tai-Woo Chiu, C. P. Hartman, and D. A. Thomas, *Intern. J. Polymeric Mater.* **1**, 331 (1972).

41. G. M. Yenwo, J. A. Manson, J. Pulido, L. H. Sperling, A. Conde, and N. Devia-Manjarres, *J. Appl. Polym. Sci.* **21**, 1531 (1977).

42. G. M. Yenwo, L. H. Sperling, J. A. Manson, and A. Conde, in *Chemistry and Properties of Crosslinked Polymers*, S. S. Labana, ed., Academic Press, New York (1977).

43. G. Allen, M. J. Bowden, G. Lewis, D. J. Blundell, and G. M. Jeffs, *Polymer* **15**, 13 (1974).

44. H. Tompa, *Polymer Solutions*, Academic, New York (1956), Chap. 3.

45. P. J. Flory, *Principles of Polymer Chemistry*, Cornell, Ithaca, New York (1953).

46. G. W. Castellan, *Physical Chemistry*, Addison-Wesley, Reading, Massachusetts (1964), Chap. 9.

47. G. M. Yenwo, L. H. Sperling, J. Pulido, J. A. Manson, and A. Conde, *Polym. Eng. Sci.* **17**(4), 251 (1977).

48. D. L. Siegfried, J. A. Manson, and L. H. Sperling, *J. Polym. Sci. Phys. Ed.* **16**, 583 (1978).

49. E. A. Neubauer, D. A. Thomas, and L. H. Sperling, *Polymer* **19**(2), 188 (1978).

50. J. R. Millar, *J. Chem. Soc.* **263**, 1311 (1960).

51. K. Shibayama and Y. Suzuki, *Kobunshi Kagaku* **23**, 24 (1966).

52. K. Shibayama and Y. Suzuki, *Rubber Chem. Tech.* **40**, 476 (1967).

53. D. Klempner, H. L. Frisch, and K. C. Frisch, *J. Polym. Sci. A-2* **8**, 921 (1970).

54. Yu. S. Lipatov, L. M. Sergeeva, L. V. Mozzukhina, and N. P. Apukhtina, *Vysokomol. Soyed.* **A16**(10), 2290 (1974).

55. S. C. Kim, D. Klempner, K. C. Frisch, and H. L. Frisch, *Macromolecules* **9**, 263 (1976).

56. J. D. Lipko, H. F. George, D. A. Thomas, S. C. Hargest, and L. H. Sperling, *J. Appl. Polym. Sci.* **23**, 2739 (1979).

57. L. P. McMaster, *Macromolecules* **6**, 670 (1973).

58. L. P. McMaster, in *Copolymers, Polyblends, and Composites*, N. A. J. Platzer, ed., American Chemical Society, Washington, D.C. (1975).

59. A. Robard and D. Patterson, *Macromolecules* **10**, 1021 (1977).

60. D. Patterson and A. Robard, *Macromolecules* **11**, 690 (1978).

61. K. C. Frisch, D. Klempner, S. Migdal, H. L. Frisch, and H. Ghiradella, *Polym. Eng. Sci.* **14**, 76 (1974).

62. H. L. Frisch, K. C. Frisch, and D. Klempner, *Polym. Eng. Sci.* **14**, 648 (1974).

63. L. E. Nielsen, *J. Macromol. Sci.* **C3**, 69 (1969).

64. D. Klempner, H. K. Yoon, K. C. Frisch, and H. L. Frisch, in *Chemistry and Properties of Crosslinked Polymers*, S. S. Labana, ed., Academic, New York (1977).

65. C. H. Bamford, G. C. Eastmond, and D. Whittle, *Polymer* **12**, 247 (1971).

66. Yu. S. Lipatov, L. M. Sergeva, L. V. Karabonova, A. Ye. Nesterov, and T. D. Ignatova, *Vysokomol. Soyed.* **A18**(5), 1025 (1976).

67. Yu. S. Lipatov, T. S. Chramova, L. M. Sergeeva, and L. V. Karabanova, *J. Polym. Sci. Polym. Chem. Ed.* **15**, 427 (1977).

68. J. F. Beecher, L. Marker, R. D. Bradford, and S. L. Aggarwal, *J. Polym. Sci.* **26C**, 117 (1969).

69. J. M. Brown and J. E. Guillet, *Adv. Polym. Chem.* **21**, 107 (1976).

70. E. Helfand and Y. Tagami, *J. Chem. Phys.* **56**, 3592 (1972).

71. E. Helfand, *J. Chem. Phys.* **63**, 2192 (1975).

72. E. Helfand, *Macromolecules* **9**, 307 (1976).

73. T. A. Weber and E. Helfand, *Macromolecules* **9**, 311 (1976).
74. (a) S. D. Gehman, D. E. Woodford, and C. S. Wilkinson, *Ind. Eng. Chem.* **39**, 1108 (1947); (b) ASTM D 1052-58T, American Society of Testing Materials, Philadelphia (1958).
75. (a) R. F. Clash, Jr. and R. M. Berg, *Ind. Eng. Chem.* **34**, 1218 (1942); (b) ASTM D 1043–51, American Society of Testing Materials, Philadelphia (1951).
76. L. E. Nielson, *Mechanical Properties of Polymers*, Reinhold, New York (1962), Chap. 7.
77. J. K. Gillham, J. A. Benci, and A. Noshay, *J. Appl. Polym. Sci.* **18**, 951 (1974).
78. L. H. Sperling, H. F. George, V. Huelck, and D. A. Thomas, *J. Appl. Polym. Sci.* **14**, 2815 (1970).
79. P. E. Rouse, *J. Chem. Phys.* **21**, 1272 (1953).
80. F. Bueche, *J. Chem. Phys.* **22**, 603 (1954).
81. M. Takayanagi, H. Harima, and Y. Iwata, *Mem. Fac. Eng. Kyushu Univ.* **23**, 1 (1963).
82. R. T. DeHoff and F. N. Rhines, eds. *Quantitative Microscopy*, McGraw-Hill, New York (1968).
83. P. R. Scarito and L. H. Sperling, *Poly. Eng. Sci.* **19**, 297 (1979).
84. V. V. Shilov, Yu. S. Lipatov, L. V. Karabanova, and L. M. Sergeeva, *J. Polym. Sci. Polym. Chem. Ed.* **17**, 3083 (1979).
85. P. Debye, H. R. Anderson, Jr., and H. Brumberger, *J. Appl. Phys.* **28**(6), 679 (1957).
86. A. Guiner and G. Fournet, *Small Angle Scattering of X-Rays*, Wiley, New York (1955).
87. C. A. Vonk, *J. Appl. Crystallogr.* **6**, 81 (1973).
88. E. H. Kerner, *Proc. Phys. Soc.* **69**, 808 (1956).
89. Z. Hashin and S. Shtrikman, *Mech. Phys. Solids* **11**, 127 (1963).
90. W. E. A. Davies, *J. Phys. D* **4**, 318 (1971).
91. R. A. Dickie, M. F. Cheung, and S. Newman, *J. Appl. Polym. Sci.* **17**, 65 (1973); R. A. Dickie and M. F. Cheung, *J. Appl. Polym. Sci.* **17**, 79 (1973).
92. G. Allen, M. J. Bowden, S. M. Todd, D. J. Blundell, G. M. Jeffs, and W. E. A. Davies, *Polymer* **15**, 28 (1974).
93. A. A. Donatelli, L. H. Sperling, and D. A. Thomas, *Macromolecules* **9**, 671, 676 (1976).
94. D. R. Paul, in *Polymer Blends*, D. R. Paul and S. Newman, eds., Academic, New York (1978), Chap. 1.
95. G. Kraus and K. W. Rollman, in *Multicomponent Polymer Systems*, N. A. J. Platzer, ed., American Chemical Society, Washington, D.C. (1971).
96. B. Budiansky, *J. Mech. Phys. Solids* **13**, 223 (1965).
97. S. C. Kim, D. Klempner, K. C. Frisch, and H. L. Frisch, *Macromolecules* **10**, 1187 (1977).

7

ENGINEERING, MECHANICAL, AND GENERAL BEHAVIOR

7.1. INTRODUCTION

Because the varied synthetic techniques yield IPNs of such diverse properties, their engineering potential spans a broad gamut of modern technology. In general, the material's use will depend on such factors as phase continuity, mixing of the two networks, size of the domains, and the glass transitions of the component polymers. Several major classes of materials are readily identified: (1) tough and/or impact-resistant plastics, (2) reinforced elastomers, (3) noise- and vibration-damping compounds, (4) vulcanized rubber/rubber blends, (5) electrical insulators, (6) transport-selective compositions, (7) coatings and adhesives and (8) ion exchange resins.

This chapter will develop the engineering and mechanical behavior of IPNs, and the following chapter will explore the more industrially oriented aspects.

7.2. FILLED INTERPENETRATING POLYMER NETWORKS

7.2.1. Density and Crosslink Levels

While IPNs themselves constitute a polymer/polymer composite, many engineering applications will require greater or lesser quantities of other components, such as inorganic fillers. Several problems regarding the location of polymer II in an IPN have already been touched upon, as well as the interfacial mixing characteristics. However, the presence of a third phase introduces new complications.

Lipatov and co-workers[1-3] synthesized sequential IPNs containing a reinforcing silica filler as follows: Polyurethane samples based on oligo-ethylene glycol adipate and tolylene diisocyanate were prepared with various levels of Aerosil, a silica filler. Trimethylolpropane was used as the

167

crosslinking agent. The filler was blended in before curing. A styrene–divinyl benzene solution was swollen into the filled polyurethane for a predetermined period of time, along with the initiator, benzoyl peroxide. After polymerizing at 60°C for five days, samples with various weight ratios of polymer II to polymer I, w_2/w_1, were obtained.

Both basic scientific information and useful engineering data can be obtained from the introduction of an inorganic filler. For example, where will the styrene go in the above material? Will it avoid the filler surface, or tend to bind to it? How will the density, crosslink level, etc. be affected? Engineering questions relate to the changes in damping characteristics, tensile strength, and resistance to swelling. The data in Table 6.4 already showed the change in density, ρ, in relation to the weight ratio of the two networks, the crosslinking level of network II, and the amount of filler in network I.[1] Along with the experimental values of ρ, the theoretical values were calculated on the basis of an additivity scheme. As may be observed in Table 6.4, the experimental densities are intermediate between the densities of the individual networks. In some cases, particularly when w_2/w_1 is low, the experimental value of the density, ρ_{exp}, is lower than the theoretical density, ρ_{theor}. As the ratio w_2/w_1 and/or the filler content rises, ρ_{exp} first approximates ρ_{theor}, then exceeds it somewhat. Assuming the variation is real, it is easier to explain the densification than the rarefication. The filling up of submicroscopic voids is an obvious possibility. More interesting is the explanation that polymer II congregates at the polymer I/filler interface, turning a rarefaction[4] into a densification.

The Flory–Rehner equilibrium swelling method (see Section 4.2.2)[5] was used to evaluate the change in crosslink density of the IPNs,[1] using dioxane as the swelling agent. The Flory interaction parameter χ was found to be approximately 0.30 for dioxane and either polymer, simplifying the analysis. Table 6.4 gives the results for the effective crosslink densities, ν_e/V, where V is the polymer volume in cm^3. Table 6.4 also shows the theoretical values of ν_e/V, calculated in accordance with the law of additivity.

Most unexpectedly, the value for $(\nu_e/V)_{exp}$ is lower than $(\nu_e/V)_{theor}$ in every case. Further, the divergence between the experimental and theoretical values of ν_e/V increases with increasing levels of crosslinking in network II, and with increasing values of w_2/w_1.

An explanation of these results has been offered by Siegfried et al.[6] and Hargest et al.[7] in terms of the Thiele–Cohen swelling equation,[8] and the usual dominance of network I over network II (see Chapter 4). A clue may be seen in that the polyurethane network I has a lower crosslink density than the polystyrene network II. If network I dominates, the behavior is expected to remain closer to that of the polyurethane network than expected from additivity of properties.

7.2.2. Dielectric Loss Measurements

While the dielectric properties of a polymer system clearly have engineering value, changes in dielectric behavior with composition yield fundamental information. The behavior of the dielectric loss factor, tan δ, for the polyurethane/polystyrene sequential IPN was investigated by Lipatov *et al.*[1] Employing a frequency of 300 Hz, they studied the effect of filler over the temperature range -130 to $20°C$. In this temperature range, the polystyrene does not exhibit a maximum in tan δ; however, the polyurethane does, at a temperature just below $0°C$. Figure 7.1 shows the effect of Aerosil content on the loss peak. For convenience, the maxima in tan δ are collected in Table 7.1.

It is seen that tan δ_{max}, corresponding to the dipole segmental losses of the crosslinked polyurethane, first appears at $-10°C$ in the absence of filler, and shifts upwards to $0°C$ at the highest level of filler. There is also an upward shift with increasing levels of DVB in the polystyrene. The greater level of crosslinking (3% vs 1%) in the polystyrene undoubtedly forces an increase in the degree of mixing.

More remarkable, there is a decided drop in the height of the maximum in tan δ (Figure 7.1) as the amount of filler is increased, which points to a decreased mobility of the chains in network I. The slightly increasing density over the arithmetic average for the highly filled systems, in combination with the reduced mobility of the polyurethane chains, suggests that the polystyrene exerts hydrostatic pressure on the polyurethane, probably in connection with the filler. This plus other data (below) suggests that the

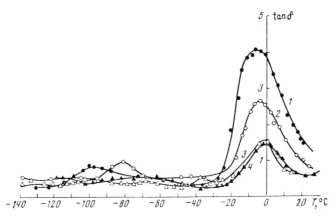

Figure 7.1. Dependence of tan δ on temperature for sequential polyurethane/polystyrene IPNs with Aerosil contents of 0%, (1) 0.5%, (2) 1%, (3) and 5% (4) with $w_2/w_1 = 0.5$ (1–3) and 0.7.[1]

Table 7.1. Dependence of the Temperature of Tan δ_{max} for
Polyurethane/Polystyrene IPNs on Filler Content in Network I
and in the IPN When $w_2/w_1 = 0.5$ [1]

Filler %	Network I	T (tan δ_{max}), °C	
		IPN	
		1% DVB	3% DVB
0.0	−10	−11	−5
0.5	−8	−9	−3
1.0	−5	−6	−1
5.0	−4	−2	0

polystyrene polymerization locus may be concentrated on the polymer
I/filler interface.

From an applications point of view, it may be that the filler yields an
unexpected advantage: reduction of dielectric heating, and concomitant
reduction in power loss. As the tan δ peak is reduced, less energy would be
transferred from an AC source to the polymer in the range of near −10 to
0°C.

7.3. ULTIMATE BEHAVIOR

When searching for improved behavior in materials, the investigator is
frequently looking for synergisms, i.e., where specific interactions lead the
properties of the whole to be greater than the arithmetic average of its parts.
Of course, such a statement implies that some materials will be average, and
others will have even below average behavior. (Too often investigators are
confronted with the latter situation!)

7.3.1. Latex IENs

Klempner et al.[9] encountered an IEN pair (see Section 5.3) which
exhibited either synergistic or nonsynergistic tensile strengths, depending on
composition (see Figure 7.2). The polymer pair was a poly(urethene-urea),
U-1033, and a polyacrylate, H-138. The variation of tensile strength with
concentrations of polyacrylate shows very interesting and unpredicted
behavior, with both a minimum and a maximum in tensile strength being
observed. As polyacrylate is added to the poly(urethane-urea), the tensile
strength initially decreases, reaching a minimum near 20% polyacrylate.
The authors[9] attribute the minimum to an initial decrease in hydrogen

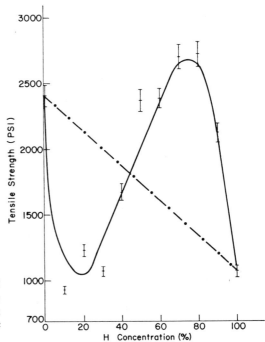

Figure 7.2. Tensile strength of a polyurethane/polyacrylate IEN: (———) experimental curve; (— · —) computed (arithmetic mean). H = polyacrylate.[9]

bonding within the poly(urethane-urea), polyacrylate acting essentially as a plasticizer. Interestingly enough, this implies no phase separation in this concentration range, see Figure 6.7.

At about 75% polyacrylate, a maximum is reached, which is actually above the tensile strength of either component. Klempner *et al.*[9] attribute this maximum to a maximum in chain entanglement, leading to an increase in physical crosslinking between networks (see Chapter 4). Sidestepping this argument for the moment, it may be noted that the T_g for the polyurethane is about −35°C and that of the polyacrylate is about +30°C (see Figure 6.28). At about 75% polyacrylate, from an empirical point of view the material may be behaving as a rubber-toughened plastic. According to Figure 6.7, this composition has well-defined, dispersed urethane rubber domains.

The effects of phase inversion may also be noted with this system. Figure 7.3[9] illustrates the elongation of these materials. If the morphological analysis of Figure 6.7 is correct, the drop off in elongations at about 80% polyacrylate may be due to the loss of phase continuity of the elastomer. This explanation is not completely satisfying, however, because polyacrylate, the "plastic" portion, is continuous over most of the composition range. Alternately, the low T_g of polyacrylate may have an

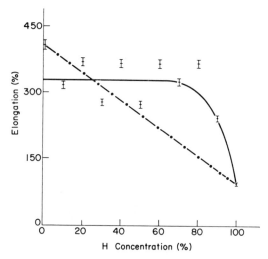

Figure 7.3. Elongation vs. polyacrylate (H) concentration: (———) experimental curve; (— · —) computed (arithmetic mean).[9]

important role as an extensible material. Additional engineering properties are pointed out in Table 8.1, example No. 10.

7.3.2. Sequential IPNs

A number of IPN compositions exhibit considerable toughness, as measured by stress–strain curves or impact strength. The main features required for IPNs are similar to those needed by graft copolymers, except that IPNs can also be utilized as tough elastomers or leathers. In both cases, the elastomeric phase domains should be in the range of 500–5000 Å in size, with the T_g of the elastomeric phase being below −40°C. Manson and Sperling[10] and Bucknall[11] review the molecular and morphological requirements in some detail.

Because of the crosslinking in polymer I, phase domains of the sequential IPNs are usually in the lower limit of the useful size range, i.e., rarely above 1000 Å. This contrasts with the usual case for graft copolymers and polymer blends, where the domain size range is often above the maximum for optimum toughness. Because of the probable dual-phase continuity, the cellular-type morphology pervades the entire structure, which is restricted to the rubber regions only in the graft copolymers, and is totally absent in the blends.

One advantage of dual-phase continuity, in tandem with double-network formation, is a degree of imposed compatibility. With such interpenetrating phases, the probability of interfacial debonding under strain is reduced. Thus dual-phase continuity in an IPN behaves somewhat like graft sites in bonding the structure together.

The continuous cellular-type structure does contribute to the toughening process, as illustrated in Figure 7.4 for SBR/PS sequential IPNs.[12] Impact values of about 5 ft-lb/in. of notch were attainable with 20–25% SBR elastomer.[12] This value contrasts with values of 1.3–2.5 ft-lb/in. of notch usually encountered with high-impact polystyrene.

Scanning electron microscopy provides insight into the actual mechanism of failure. In order to observe the effects of loading rate on fracture, tensile test specimens were compared to the impact loaded specimens. Figure 7.5[12] shows a scanning electron micrograph of the behavior of a semi-IPN of the second kind in fracture. (A corresponding transmission electron micrograph is found in Figure 2.3, which shows the polystyrene as the discontinuous phase.) The discontinuous polystyrene domains are most evident in Figure 7.5c, where slow loading has allowed a greater relaxation of the bulk material, and consequently, a greater development of the failure mechanism prior to fracture. It appears that the fracture lines followed the rubber/plastic interface because the PS domains are intact. By contrast, Figure 7.5d shows a gross deformation of the polystyrene phase. Although these semi-II IPN compositions are not particularly tough, they illustrate the molecular and supermolecular energy-absorbing deformations most clearly. One may imagine that forcing the hard polystyrene domains to deform before the material fails significantly improves the total energy absorption.

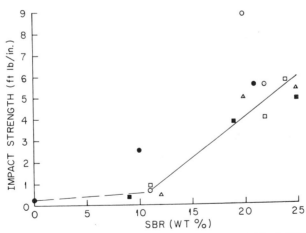

Figure 7.4. Impact strength as a function of composition and crosslinking in PS. In all samples SBR crosslinked with 0.10% Dicup. Series 3 and 12: PS is not crosslinked. Series 4: PS crosslinked with 1% DVB. Series 5 and 13: PS crosslinked with 2% DVB.[12]

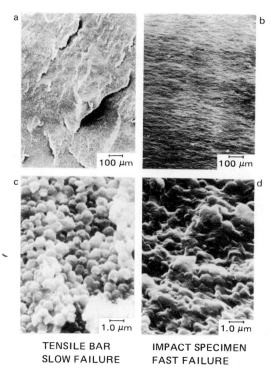

Figure 7.5. Micrographs of the fracture surfaces of test specimens from series 2: (a) Low magnification of tensile test 20% SBR (5% S)/80% PS with SBR not crosslinked and PS crosslinked by 2% DVB; (b) low magnification of impact test 20% SBR (5% S)/80% PS with SBR not crosslinked and PS crosslinked by 2% DVB; (c) high magnification of tensile test, 20% SBR (5% S)/80% PS with SBR not crosslinked and PS cross-linked by 2% DVB; (d) high magnification of impact test, 20% SBR (5% S)/80% PS with SBR not crosslinked and PS crosslinked by 2% DVB.[12]

7.3.3. Simultaneous Interpenetrating Networks

7.3.3.1. Stress-Strain Studies

Of the several methods of preparing double-network compositions, the synthesis of SINs probably offers the greatest versatility. One outstanding advantage is improved processibility.

In order to evaluate the effect of the several variables in an epoxy/poly(n-butyl acrylate) SIN, a computer program was used to assist in the analysis.[13] The experimental design involved a rotatable central composite design.[14] The three independent variables chosen were epoxy prereaction time, initiator level (for the acrylic), and percent diethylene glycol dimethacrylate (DEGDM) crosslinker for the acrylic. Sixteen compositions were chosen, as illustrated in Figure 7.6.[13] The center of the experimental design was chosen as the point thought to give simultaneous gelation with both reactions proceeding at a moderate rate.[15]

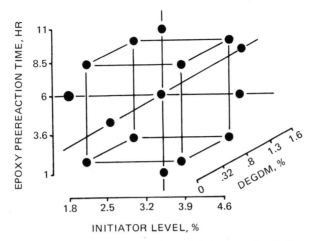

Figure 7.6. Rotatable central composite experimental design, showing the three independent SIN variables.[13]

Typical stress–strain curves are shown in Figure 7.7.[13] Illustrated are the epoxy homopolymer (Shell 828, cured with 31.1 parts of phthalic acid anhydride per hundred parts of epoxy), and three SIN compositions having epoxy prereaction time as the variable. Each of the SINs is seen to be significantly tougher than the parent epoxy. The absolute tensile strength is higher, and the area under the stress–strain curve, a measure of the work to

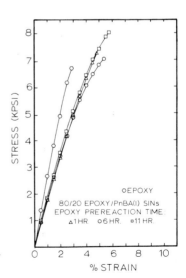

Figure 7.7. Stress–strain curves for an epoxy homopolymer and three epoxy/PnBA SINs.[13]

break, is increased. The initial slope, indicative of Young's modulus, is lower, showing that the rubber phase softens the materials somewhat.

With the aid of the computer program, contour maps of several properties were prepared. Figure 7.8[13] illustrates one such map. The contours indicate lines of constant tensile strength. The diagonal from the lower left to the upper right in Figure 7.8 roughly yields a line of simultaneous gelation conditions. Most unexpectedly, higher tensile strengths were observed in the upper left and lower right quadrants of the map, on either side of the simultaneous reaction locus. Thus regions of simultaneous gelation yielded the weakest materials. This was attributed to the relatively fine phase domain structure attained at simultaneous gelation conditions.[15] When one polymer or the other was allowed to polymerize faster, to produce nonsimultaneous gelation conditions, the phase domains were larger and the materials consequently stronger.

In general, the epoxy/acrylic SIN exhibited poor compatibility, which was subsequently improved by grafting the two polymers together with glycidyl methacrylate.[16] It was shown that as the grafting level was increased, the T_g of the epoxy phase decreased significantly and room temperature stress–strain curves reflected a significant softening of the product.

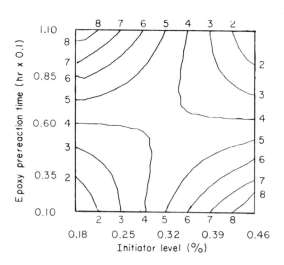

Figure 7.8. A contour map of tensile strength for the epoxy/poly(n-butyl acrylate) SIN pair. The crosslinker for the poly(n-butyl acrylate), DEGDM, was held at the constant level of 0.80%.[13] The contour values (in psi, all $\times 10^3$) for the contour symbols are as follows: symbol 1, 6.04; 2, 6.46; 3, 6.89; 4, 7.31; 5, 7.74; 6, 8.16; 7, 8.59; 8, 9.44.

7.3.3.2. Hardness

Kim *et al.*[17] studied the engineering urethane/poly(methyl methacrylate) SINs, linear bl⟨ materials (one polymer crosslinked). The PU/PMM better compatibility than the epoxy/PnBA systen simple property, though of significant usefulness, tne ⌄⌄ the materials was evaluated. The indentation hardness, which ⌐⌐⌐⌐ resistance to local deformation, is a complex property related to the modulus, strength, elasticity, and plasticity.[18] For the case of elastic materials, the depth of penetration, h, of a spherical indentor of radius R is related to Young's modulus, E, Poisson's ratio, ν_ρ, and the total force F by

$$E = \frac{3}{4} \frac{1 - \nu_\rho^2}{h^{3/2}} FR^{-1/2} \qquad (7.1)$$

From a five-senses point of view, a ball indentor measures about the same properties as detected by pressing on a solid sample with one's thumb; however, quantitative engineering values are obtained with proper instrumentation.

Typical results are shown in Figure 7.9.[17] The high Shore A hardness values obtained for low PU concentrations are typical of plastic compositions. The plot also reflects the phase inversion process; the hardness rapidly decreases in the 60–70% polyurethane concentration range.

* Again, the Frisch team has adopted the term "pseudo" where Sperling *et al.* use "semi," and the term "linear blend" where Sperling *et al.* use "chemical blend."

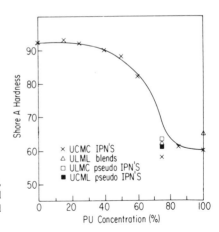

Figure 7.9. Hardness vs. composition for SINs, pseudo-IPNs, and linear blends, prepared from polyurethane and poly(methyl methacrylate).[17]

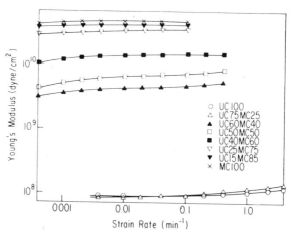

Figure 7.10. Young's modulus vs. strain rate for the polyurethane/poly(methyl methacrylate) SINs at 23°C. Note phase inversion, as indicated by the drop in modulus.[20]

A basic measure of hàrdness, of course, is the modulus of the material. While the modulus of an ideal elastic body remains independent of the conditions of measurement, the modulus of a real viscoelastic body varies as a function of time and temperature. The modulus decrease with increasing temperature was illustrated in Figures 6.26 and 6.28. Through the application of the concepts embodied in the time–temperature superposition principle,[19] the modulus would be expected to increase with increasing strain rate.

This effect was studied by Kim et al.[20] (see Figure 7.10). On increasing the strain rate by a factor of 1000, Young's modulus is seen to increase by about 50%. The time–temperature superposition principle states that an increase of the rate of an experiment by a factor of 10 is equivalent to lowering the temperature by 6–7°C.* That greater modulus changes were not observed is due to the fact that test conditions were selected either far above or far below the two glass transitions.

Of course, the change in modulus from the high group in Figure 7.10 to the low group is indicative of a phase inversion. The two lower curves are either 75 or 100% elastomer, and have the elastomer as the continuous phase.

Corresponding tensile and tear strength results are shown in Table 7.2.[17] The ultimate tensile strength reached a maximum with the UC85MC15 SIN, which shows significant synergistic behavior. The morphological studies on this material showed that the PU phase is

* The T_g of a material increases 6–7°C per decade of experimental rate increase.

continuous at this concentration, leading to a reinforced elastomer material. The SINs shown in Table 7.2 are improved over the corresponding blends and semi-SINs. This reflects the interpenetration effect, since the domain sizes of the SINs were significantly smaller than the blend and semi-SINs. The authors also pointed out that the leathery samples, UC40MC60 and UC60MC40, showed stress whitening during elongation, an indication of the development of internal crazing as a toughening mode.

Belonovskaya et al.[21] made a detailed study of their semi-SIN materials, based on isocyanates. (See Section 5.5.2 for synthetic details.) Table 7.3 shows tensile strength, modulus, and impact strength for a number of polymer pairs. The polypropylene sulfide PPS/TDI ratio is varied in Figure 7.11. It should be noted that the polyisocyanate is very rigid in structure.

7.3.3.3. Thermal Degradation

The thermogravimetric properties reported by Kim et al.[17] exhibit a most unusual synergism. While some compositions do fall intermediate between the PU and PMMA components (see Figure 7.12), the UC75MC25 and related compositions show a significantly greater temperature resistance than the homopolymers, as shown in Figure 7.13.[17]

Table 7.2. Polyurethane/Poly(methyl methacrylate) SIN Tensile and Tear Strengths[a][17]

Composition	Ultimate tensile strength, psi	Elongation at break %	Tear resistance, lb/in.
Homopolymers			
UC100[b]	5159	780	252
UL100	5629	815	339
MC100	4106	4.9	—
SINs			
UC85MC15	6096	767	294
UC75MC25	5127	833	355
UC60MC40	3265	300	381
UC40MC60	2592	43	—
Semi-SINs			
UC75ML25	4072	728	364
UL75MC25	4157	749	369
Chemical blends			
UL75ML25	4280	853	341
Average error range	500	71	16

[a] Cross-head speed, 2 in./min.
[b] See Table 5.6 for designations.

Table 7.3. Some Physicomechanical Properties of Semi-SINs Based on
Diisocyanates and Polar Monomers Prepared under Optimum Conditions[21]

No.	$M_1 : M_2$ $1:2$	T_{soft} (θ_s), °C	Tensile strength (δ_B), kg/cm^2	Elongation (ε_B), %	Elastic modulus (E), kg/cm^2	Impact strength (η), kg/cm^2
1	TDI:PPS	240–250	430–470	4	15,000	13–14
2	HDI:ES	250–260	600–670	6–9	17,000–18,000	—
3	TDI:MMA	230–250	600–800	3–3.5	20,000–22,000	5–6
4	TDI:Ma	200	560–600	3–3.6	22,000–23,000	22–28
5	TDI:ACR	200–250	400	2–2.8	17,000–18,000	—
6	TDI:AN	>350	450–500	3–3.5	~17,000	—

The clustering of the PU/PMMA curves in Figure 7.13 indicates that there is no significant difference among them in degradation behavior, and that the enhancement of the weight retention is not due directly to inter= penetration. A likely mechanism has been suggested by Belyakov *et al.*,[22] who suggested a synergistic radical mechanism. The PMMA degrades via a stepwise unzipping process first. The improvement in thermal degradation is explained in terms of the more easily degradable polymer-absorbing free radicals, stabilizing and protecting the less easily degradable polymer, in this case the PU network. Thus the PMMA and its degradation products act as free-radical scavengers for the PU. This mechanism requires a close jux- taposition of the two polymers, which may be enhanced by IPN formation.

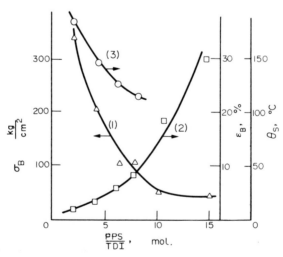

Figure 7.11. Dependence of some thermomechanical properties of semi-SINs on PPS/TDI ratio. (1) Tensile strength; (2) elongation; (3) softening temperature.[21]

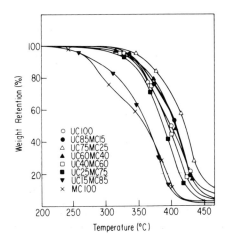

Figure 7.12. TGA thermograms for polyurethane/poly(methyl methacrylate) SINs.[17]

Kim *et al.* reported similar thermogravimetric analysis (TGA) results on polyurethane/polystyrene SINs.[23]

Both the elastomeric and the plastic castor oil/polystyrene SINs exhibit significant toughening.[24–27] The synthesis of these materials is discussed in Sect on 5.5.2. Figure 7.14[26] shows the stress–strain behavior of COPEN, COPEUN, and COPUN single networks, and their PSN-reinforced SINs. In each case, the PSN tends to reinforce the castor oil elastomer, in a manner similar to carbon black reinforcing SBR. The modulus, strain to break, and work to break are vastly increased. The pertinent values are summarized in Table 7.4.[26]

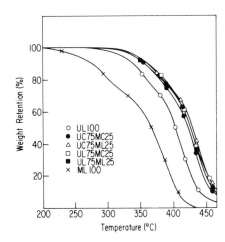

Figure 7.13. TGA thermograms for 75/25 (PU/PMMA) SIN, semi-SINs, and chemical blends.[17]

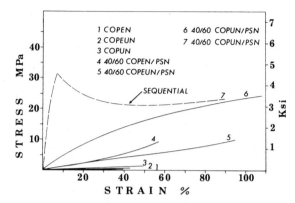

Figure 7.14. Stress–strain curves for SINs containing 40% castor oil elastomer. Discontinuous curve adapted from the sequential IPN synthesis.[26]

Also shown in Figure 7.14 is a stress–strain curve of a castor oil–urethane/polystyrene sequential IPN[28] (see curve 7). The SIN illustrated in curve 6 has the same stoichiometry; the two materials differ only in the method of preparation. The yield point shown in curve 7 suggests that the polystyrene in the sequential IPN has a significantly greater degree of phase continuity than the SIN counterpart. This is borne out by electron microscope studies (compare Figures 6.6 and 6.13).

The tensile properties of the corresponding plastic SINs are shown in Table 7.5.[26] The Izod impact strength for the COPEN/PSN 10/90 plastic is 67.8 J/m, an improvement of a factor of 3 over the polystyrene homopolymer.

Table 7.4. Tensile Properties of Castor Oil/Polystyrene SINs at Ambient Conditions: Elastomeric Compositions[26]

Composition	Tensile strength,[a] MPa	% Strain at break	Elastic modulus, MPa	Fracture energy,[b] J/m^3
40/60 COPEN/PSN	9.2	57	13.1	2.17
40/60 COPEUN/PSN	9.8	95	24.8	4.47
40/60 COPUN/PSN	24.1	108	65.5	16.85
COPEN	0.35	39	1.5	0.07
COPEUN	0.70	43	2.3	0.15
COPUN	0.97	58	2.8	0.29

[a] Samples tested at 2.11×10^{-5} m/sec (ASTM D 1708).
[b] Calculated based on the area under the stress–strain curves.

Table 7.5. Tensile Properties of SINs Containing 10% Castor Oil Elastomer at Ambient Conditions[26]

| Composition | Tensile stress[a] | | % Strain at break | Elastic modulus, MPa | Fracture energy,[b] J/m^3 | Izod impact strength, J/m |
	at yield, MPa	at break, MPa				
PSN	—	46.1	2.2	2360	0.58	13.3
COPEN/PSN	31.1	37.0	16.0	1520	5.37	67.8
COPEUN/PSN	22.1	25.5	18.5	1090	3.46	44.4
COPUN/PSN	37.3	36.7	6.4	1680	2.01	24.6

[a] Samples tested at 2.11×10^{-5} m/sec (ASTM D 1708).
[b] Calculation based on the area under the stress–strain curves.

7.3.3.4. Lap-Shear and Peel Strength

For prospective adhesive or coating materials, lap-shear and peel experiments provide an analysis of necessary but quite contrasting properties. Peel strength always measures the interfacial composite bond strength. Lap-shear studies usually measure the internal or cohesive strength of the polymer subjected to shearing stresses. While lap-shear experiments can result in interfacial failure, under good adhesive bonding conditions the polymer itself usually fractures.

Typical experimental designs for lap-shear tensile and peel strength studies are illustrated in Figures 7.15 and 7.16, respectively.[29] In the lap-shear tensile test, a stress–strain instrument such as an Instron pulls the two plates shown. From elasticity theory, a roughly linear relationship is expected between the logarithm of the lap-shear strength and the logarithm of the tensile strength.

Lap-shear studies for several polyurethane–epoxy SINs and semi-SINs are reported in Figure 7.17.[29] A distinct maximum in lap-shear strength is noted at midrange compositions. By contrast, the peel strength shows a

TEST SPECIMEN

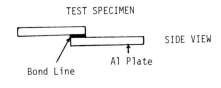

SIDE VIEW

Al Plate
Bond Line

Figure 7.15. A lap-shear tensile test specimen, according to ASTM D-3163-73.[29]

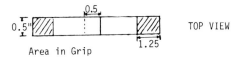

TOP VIEW

Area in Grip

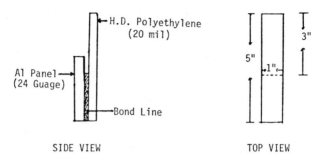

SIDE VIEW   TOP VIEW

Figure 7.16. A peel strength test specimen, according to ASTM D-903-49.[29]

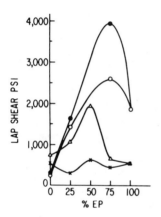

Figure 7.17. Lap-shear (psi) vs. network composition for polyurethane/epoxy SINs: SIN-I (open circles); SIN-II (triangles; semi-SIN (filled circles); SIN-III (crosses).[29]

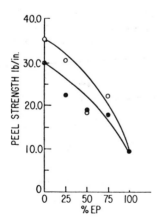

Figure 7.18. Peel strength (lb/in.) vs. network composition for polyurethane/epoxy SINs: SIN-I (open circles); semi-SIN-I (filled circles).[29]

continuous decline with increasing epoxy content, as illustrated in Figure 7.18.[29] Probably the epoxy, being a glassy polymer, cannot deform as well as the urethane under peel conditions. An alternate explanation is that the bonding characteristics of PU are better towards the polyethylene outer layer than epoxy. In either case, interpenetration as such appears not to play a decisive role in adhesion (as opposed to cohesion), at least for this system.

7.4 ELECTRICAL AND BARRIER PROPERTIES

7.4.1. Volume Resistivity

While stress–strain and impact tests are destructive in nature, electrical insulation and permeability measurements usually are not. This section will discuss the behavior of IPNs in these key engineering applications.

Kim *et al.*[30] investigated the volume resistivity of poly-urethane/PMMA SINs. The experimental data were analyzed in terms of the theoretical composite models of Maxwell[31,32] and Böttcher.[33] The Maxwell model assumes that an "average sphere" is surrounded by an annulus of the matrix material, and that the composite is composed of many equivalent sphere–annulus portions, each having the same volume resistivity as the composite. The final equation can be expressed

$$\frac{K}{K_m} = \frac{(1 - v_i)K_m + (2 + v_i)K_i}{(1 + 2v_i)K_m + 2(1 - v_i)K_i} \tag{7.2}$$

where K, K_m, and K_i are the volume resistivities of the composite, matrix, and inclusion, respectively, and v_i is the volume fraction of the inclusion, (dispersed phase). The approach by Maxwell was quite similar to Kerner's equation for the modulus behavior of composite materials. The Maxwell equation and the corresponding Kerner equations yield both an upper and a lower bound for materials, depending on which phase is assumed continuous.

Böttcher[33] attempted to interpolate between the upper and lower bounds of the Maxwell equation. He assumed two co-continuous phases, in a manner similar to Budiansky[34], see Section 6.9, and especially (6.55). The Böttcher equation may be written

$$\frac{v_1}{1 + 2/3(K_1/K - 1)} + \frac{v_2}{1 + 2/3(K_2/K - 1)} = 1 \tag{7.3}$$

where K, K_1, and K_2 constitute the volume resistivities of the composites and phases 1 and 2, respectively.

Note that like the Budiansky equation, the Böttcher equation is symmetrical in form. It has been suggested that both Budiansky and the Böttcher approaches predict the composite property especially well when the system undergoes a phase inversion or exhibits dual-phase continuity.

The experimentally measured DC volume resistivities of the SINs are shown in Figure 7.19. The calculated volume resistivities of the Maxwell model, assuming the urethane phase continuous, or assuming the PMMA phase continuous, are shown. The volume resistivities based on the Böttcher model are also shown in Figure 7.19. In Figure 7.19 it can be seen that the fit of the Böttcher model is superior to either Maxwell model. This is in line with the morphological studies on these same materials (Section 6.4.2), which show that a phase inversion occurs at an intermediate position.

7.4.2. Barrier Properties

From a fundamental point of view, different underlying morphologies are sensed differently for various properties. Thus mechanical measurements might be more responsive to, say, phase continuity, while thermal measurements might be more sensitive to the extent of actual molecular mutual solution. It is therefore interesting to bring to bear barrier properties as another class of molecular probes by which an IPN can be studied. The behavior of permeating small molecules can yield much information

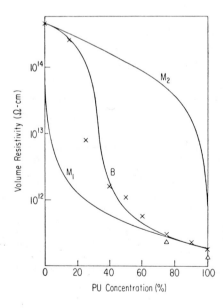

Figure 7.19. Volume resistivity of polyurethane–poly(methyl methacrylate) SINs (×) and chemical blends (△). Solid lines are the calculated curves based on the Maxwell equation [M_1 assuming the polyurethane as the matrix, M_2 assuming the poly(methyl methacrylate) as the matrix], and on the Böttcher model (B).[30]

regarding phase continuity, tortuosity, etc., provided that the penetrant permeabilities of the pure components are sufficiently different and well known.

From an applications point of view, IPNs are potentially useful as coating materials, packaging, etc. Frisch *et al.*[29,35] recently studied the barrier behavior of polyurethane/epoxy SINs. Figure 7.20 reports the permeability, diffusion, and sorption coefficients of water vapor vs polymer I/polymer II overall composition. Interestingly enough, the permeability data exhibit an S-shaped curve as the composition is altered. Frisch *et al.*[29,35] reported that the permeability, P, must lie between an upper and lower bound:

$$\frac{1}{v_1/P_1 + v_2/P_2} \le P \le v_1P_1 + v_2P_2 \tag{7.4}$$

where P_1 and P_2 and v_1 and v_2 are the permeabilities and volume fractions of polymers 1 and 2, respectively. While the data in Figure 7.20 satisfy the above inequality, it appears that a treatment similar to Böttcher's or Budiansky's would be more specific. The inflection point seems related to the point of phase inversion. In the vicinity of the inflection point, the quantities are changing in the opposite directions. The diffusivity, D, decreases by a factor of about 7, while sorption, S, increases by about 2.5 in

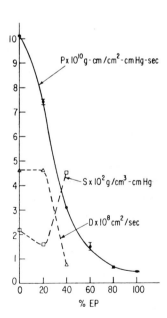

Figure 7.20. Water vapor permeability, diffusion, and sorption coefficients vs. network composition for a SIN based on a polyurethane and an epoxy.[35]

the range of epoxy (EP) concentration of 20–40%. Of course, the several terms are related by

$$D \cdot S = P \qquad (7.5)$$

where P represents the permeability.

More recently, Frisch and Frisch[35] reported on the toluene vapor transmission through an SIN film based on poly(1,4-oxybutylene) glycol, MW: 1000, and 4,4'-methylene bis(cyclohexyl isocyanate) (H_{12}MDI), and epoxy resin. The coefficient of diffusion decreased, as did the permeability, as the epoxy content was increased.

7.4.3. Critical Surface Tension

Frisch and Frisch[35] also measured the critical surface tension on several epoxy/polyurethane SINs. Water–methanol and methanol–ethylene glycol mixtures were employed, using the advancing contact angle method. Interestingly, at the network compositions where the SIN samples possessed maxima in their ultimate mechanical properties, such as lap-shear strength (see Figure 7.17), the critical surface tensions exhibited pronounced minima.

Frisch and Frisch[35] speculated that the minimum in surface tension might be due to a large entropic contribution to the reversible work of wetting. This, in turn, may have been caused by an elastic straining of the immediate surface layers near a critical point of inversion. One of the network components may have been leaving the interface, and the other migrating there at the minimum.

7.5. IONICALLY CHARGED IPNs

In simple homopolymers and networks, ionic charges may be introduced for a variety of reasons. For example, introducing about 5% of sodium methacrylic acid into polyethylene produces ionomers,[36,37] which are clear, tough plastics. When the T_g of the polymer is below ambient, materials known as carboxylic rubbers are formed. Anionic and cationic ion exchange resins have been known for a long time.[38,39]

Three engineering uses have been proposed for ionic IPNs: (1) cationic/anionic ion exchange resins, (2) piezodialysis membranes, and (3) thermoplastic IPNs. The last item is discussed briefly in Sections 5.7 and 8.7. The first two will be briefly treated here.

7.5.1. Cationic/Anionic Ion Exchange Resins

The first report of anionic/cationic IPNs for ion exchange resins is due to Solt.[40] A brief quote from one of his syntheses follows[40]:

> As an example of such a process, beads or other particles of a copolymer of styrene and divinyl benzene may be caused to react with chloromethyl ether in the presence of anhydrous aluminium chloride to give a product containing chlormethyl groups. At this stage the particles have not ion-exchange properties. They are washed dried, and then immersed in a mixture of styrene and divinyl benzene containing benzoyl peroxide as a catalyst. The particles are allowed to remain in this mixture until they cease to imbibe it, this step taking say 30 minutes, and then are separated from any residual mixture and subjected to the usual polymerisation treatment, which is generally carried on with the particles in suspension in water or other liquid. When the imbibed monomeric liquid has been polymerised, it is subjected to two successive chemical treatments. The first of these comprises treatment with trimethylamine to give quaternary ammonium groups by reaction with the chlormethyl groups in the original polymer. The second treatment is effected with hot concentrated sulphuric acid which introduces sulphonic acid groups into the polymer formed in the second polymerisation. These sulphonic acid groups give the product cation-exchange properties, which are modified by the existence of the quarternary ammonium groups.

It is especially interesting to note that this patent, issued in 1955, was written before the term "interpenetrating polymer network" was coined (see Section 1.3).

More recently, a series of IPNs was made where polymer I has a reticulated or macroporous form.[41-47] In one study[46] an IPN of poly(styrene-co-DVB)/poly(methacrylic acid-co-DVB) was made by the following route: styrene, divinyl benzene (3–20%), and paraffin oil were mixed. During polymerization, the polystyrene network precipitated to form a separate phase, yielding a porosity in the range 0.4–0.6. The crosslinking must be above 3% to prevent a physical collapse of the structure.[44] While the exact mechanism is not yet understood, it is probably important to form a continuous network before phase separation ensues. Pore diameters ranged from 2 to 4 μm, as measured by high-pressure mercury porosimetry. For this technique to work, it must be added, the pore structure must possess some degree of continuity. Thus a structure resembling Swiss cheese might be envisioned. The optimal parameters associated with network II and the final IPN composition are shown in Tables 7.6 and 7.7. The polymerization of network II takes place within the pores of network I, which undergoes a minimum of swelling. Thus network I remains in an unstrained circumstance.

In another study, Trochimczuk and co-workers[44-47] swelled polyethylene with styrene/divinyl benzene mixtures. In the range of 2–20% DVB, the polystyrene network formed a continuous phase,[47] leading to a type of dual-phase continuity. The films were then treated with methyl-

Table 7.6. Optimal Parameters of Network II Formation[46]

Impregnation ratio	14 cm^3 of monomer mixture per gram of polymer I
DVB concentration	1.5% by weight (related to monomers)
Inert diluent/n-heptane concentration	13.3% by volume
Initiator/benzoyl peroxide concentration	2% by weight
Temperature	95°C
Time of reaction	90 min

chloromethyl ether in the presence of $SnCl_4$, $ZnCl_2$, and $AlCl_3$. The chloromethylation prepares the materials for ion exchange reactions.[40] However, significant amounts of polyethylene may be crosslinked through chloromethylation[44].

$$
\begin{array}{c}
| \\
CH_2 \\
| \\
-CH_2-CH-CH_2-
\end{array}
\xrightarrow{CH_3-O-CH_2Cl}
\begin{array}{c}
| \\
CH_2 \\
| \\
-CH_2-C-CH_2- \\
| \\
CH_2Cl
\end{array}
+ CH_3OH
\qquad (7.6)
$$

$$
\begin{array}{c}
| \\
CH_2 \\
| \\
-CH_2-C-CH_2- \\
| \\
CH_2Cl \\
+ \\
-CH_2-CH-CH_2- \\
| \\
CH_2 \\
|
\end{array}
\xrightarrow{cat}
\begin{array}{c}
| \\
CH_2 \\
| \\
-CH_2-C-CH_2- \\
| \\
CH_2 \\
| \\
-CH_2-C-CH_2- \\
| \\
CH_2 \\
|
\end{array}
+ HCl
\qquad (7.7)
$$

or bound to aromatic rings through CH_2Cl groups,

$$
\begin{array}{c}
-CH_2-CH- \\
\bigcirc \\
CH_2Cl \\
+ \\
-CH_2-CH-CH_2- \\
| \\
CH_2 \\
|
\end{array}
\xrightarrow{cat}
\begin{array}{c}
-CH_2-CH- \\
\bigcirc \\
CH_2 \\
| \\
-CH_2-C-CH_2- \\
| \\
CH_2 \\
|
\end{array}
+ HCl
\qquad (7.8)
$$

Table 7.7. Composition of Reticulated IPNs by Weight[46]

Fraction of Polymer I (Styrene–DVB)	0.4
Fraction of Polymer II (Methacrylic Acid and DVB)	0.6
Cation exchange capacity	6 ± 0.5 milliequivalents/g of IPNs

It may be surmised that the polyethylene serves to mechanically strengthen the final ion exchange resin by providing mechanical support.

The concept of using a porous or macroreticular network I to provide mechanical strength to an ion exchange resin was advanced by Barrett and Clemens,[48] who coined the term "hybrid resins" to distinguish their materials from the prior art. One of the networks, either I or II, was densely crosslinked, and the other lightly crosslinked. Both homo-IPN- and hetero-IPN-type structures were prepared. One advantage imparted by such a composition is that the densely crosslinked network provides mechanical strength, while the lightly crosslinked network swells greatly, providing rapid ion exchange capability. Both strong base and strong acid ion exchange resins were made this way,[48] as well as materials containing acrylic acid or 2-vinyl pyridine. Crosslinking was usually done with divinyl benzene.

Amphoteric ion exchange resins containing a weak acid and a weak base, in IPN form, were described by Hatch.[49] Polymer network I, in microbead form (0.1–$10\ \mu$m), prepared by emulsion polymerization or related techniques, is dispersed in monomer mix II, followed by a second polymerization.

The sequence of reactions to form a microbead/matrix resin may be delineated as follows[49]:

1. A crosslinked weak acid or weak base ion exchange resin in microbead form was prepared with an average particle size of about 0.1–$10\ \mu$m and about 0.5–10% crosslinking.

2. The weak acid or weak base microbead particles were suspended in a liquid mixture of monomers which can be converted to a crosslinked matrix polymer of opposite ion exchange type.

3. The liquid monomer mixture II was polymerized to form a matrix resin having the microbead particles embedded therein.

4. The microbead/matrix resin was converted, as necessary, into active ion exchange form with a particle size of about 200–5 mesh (0.074–400 mm).

The microbead and matrix resins can be formed directly in the desired size by emulsion or suspension polymerization techniques, respectively, or as larger particles which can be subsequently converted to proper size. Also the component microbead and matrix resins can be converted to the desired weak base or weak acid form by any of a variety of reactions known for the synthesis of ion exchange resins.

If the resin has microbeads of crosslinked polyacrylic acid embedded in a matrix of poly(vinyl benzyl chloride) aminated with dimethyl amine, the amphoteric IPN can absorb salts from water in a thermally reversible manner:

$$-RCOOH + -RNR'_2 + NaCl \underset{hot}{\overset{cold}{\rightleftharpoons}} -RCOO^-Na^+ + -RNR'_2H^+Cl^-$$

$$(7.9)$$

7.5.2. Piezodialysis Membranes

Piezodialysis is a novel desalination technique in which salt is preferentially transported across the membrane and removed from a feed using pressure as the driving force.[50-52] The theory requires the membrane to consist of two continuous phases, one anionic and one cationic.[53] A suitable material should have dual-phase continuity, but with a minimum of molecular mixing between the phases. The morphological features attainable with IPN formation fit these theoretical requirements.[50,54]

Anionic/cationic IPN membranes were synthesized by sequential polymerization of crosslinked polystyrene as network I, and crosslinked poly(4-vinyl pyridine), P(4-VP), as network II.[54] Ionomeric substitution of the two networks was based on sulfonation and quaternization of the phenyl and pyridine rings, respectively. Electron microscopy involved alternate staining of the anionic and cationic phases using CsF and LiI, and showed a two-phase structure (see Figure 7.21[54]). Comparison of the two staining techniques yielded strong evidence of a positive/negative, negative/positive phase contrast, depending on the phase being stained. A phase inversion appears to occur between 50 and 80% P(4-VP), see Figure 7.21b, c, f, and g, suggesting a range of dual-phase continuity.

Compositions a and e in Figure 7.21 have only network I continuous, while compositions d and h appear phase inverted.

7.6. GRADIENT IPNs

All of the compositions reported above are uniform on a macroscopic scale through the material. IPNs can be produced, however, in which the structure or composition of the macroscopic material is not homogeneous

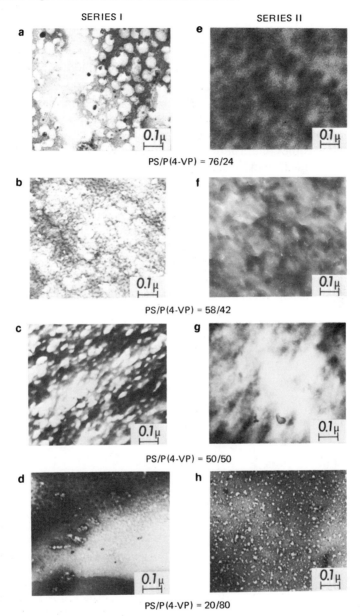

SERIES I   SERIES II

PS/P(4-VP) = 76/24

PS/P(4-VP) = 58/42

PS/P(4-VP) = 50/50

PS/P(4-VP) = 20/80

Figure 7.21. High-magnification electron micrographs of ionized IPNs PS/P(4-VP) = 76/24, 58/42, 50/50, 20/80. Series I (CsF stain) shows sulfonated PS as the dark phase, while series II (LiI stain) shows quaternized P(4-VP) as the dark phase. Comparing series I and II suggests a complementary staining. Domain sizes decrease as the P(4-VP) content increases, with a phase inversion occurring between 50 and 80% P(4-VP).[54]

throughout, but varies as a function of position. An extreme example consists of a layered or laminated material, where the gradient in composition is a discontinuous step change. Other more gradual or continuous composition gradients can easily be envisioned, e.g., linear, sigmoidal, or parabolic gradients. One potential advantage of gradual gradient IPNs is improved structural integrity in that delamination may be avoided.[55]

Gradient IPNs may easily be prepared by swelling polymer network I with monomer mix II on either one or both sides, and polymerizing before the swelling becomes uniform through diffusion.[55–59]

Sperling and Thomas[58] employed gradient IPNs to obtain a hard exterior, a soft interior, and a composition-graded intermediate zone. Such materials were useful for noise and vibration damping as explored in Section 8.8. Predecki[59] swelled hydroxyethyl methacrylate monomer mixes into silicone rubber to produce materials having hydratable surfaces. Such materials could replace surface coatings in arteriovenous shunts.

Akovali *et al*[55] prepared crosslinked polystyrene and poly(methyl methacrylate) sheets, and dipped them into acrylonitrile or methylacrylate, respectively. After various periods of time, the materials were removed and

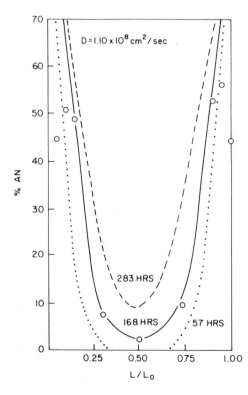

Figure 7.22. Gradient profile of PS/Grad AN prepared by immersing a polystyrene crosslinked sheet in acrylonitrile monomer at 50°C. Open circles were experimental data determined by combusion analysis. Curves were computed by Fick's equation, using a diffusion coefficient of $1.10 \times 10^8 \text{ cm}^2/\text{sec}$ for immersion times of 57, 158 and 283 hr, respectively.[55]

photopolymerized with a UV source. Figure 7.22 shows a typical gradient profile. The stress–strain behavior of PMMA/PMA IPNs, random copolymers, and gradient compositions is illustrated in Figure 7.23. The gradient IPN has the highest yield stress, and the largest area under the curve. The quantity L/L_0 is the fractional sample thickness.

Very recently, Berry et al.[56,57] prepared a series of physical/chemical IPNs and gradient IPNs based on polyether–urethane-urea (PEUU) block copolymers and acrylamide, 2-hydroxyethyl methacrylate (HEMA), or N-vinyl-2-pyrrolidone. Intended for biomedical uses, such compositions created high-strength, water-absorbing hydrogel surfaces showing good blood compatibility.

7.7. VULCANIZED RUBBER/RUBBER BLENDS

The definition of an IPN includes all cases of two crosslinked polymers, where one or both polymers are polymerized and/or crosslinked in the

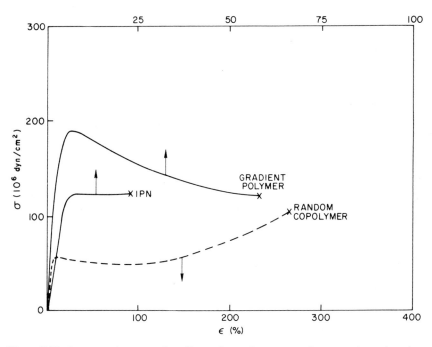

Figure 7.23. Stress–strain curves of gradient polymer, interpenetrating networks, and random copolymer of methyl methacrylate and methyl acrylate (ratio 60/40); temperature, 80°C; strain rate, 0.03 sec^{-1}. (From Akovali et al.)[55]

immediate presence of the other.* In the case of most rubber/rubber blends, especially as used in automotive tires, the polymers are mechanically blended together in the linear state, and subsequently crosslinked via sulfur vulcanization.[60-63] Thus such materials constitute a specialized type of IPN.

The importance of such rubber blends/IPNs cannot be overestimated. Nearly three million metric tons of rubber are used annually in the United States.[64] Some 64% of that rubber goes into automotive tires and nearly all of that in the form of blends,[64] for example, styrene–butadiene rubber, SBR, with *cis*-polybutadiene to improve wear, flex resistance, and cut resistance in tires.[64] Typical elastomers used in passenger and truck tires includes styrene–butadiene rubber (SBR), natural rubber (NR), synthetic

---

* An interesting exception, then, is the blending together of two separately polymerized and crosslinked latexes or suspensions.

Table 7.8. Typical Tire Tread Recipes[61]

Ingredient	phr[a] Natural rubber	Synthetic	Function
Smoked sheet	100	—	Elastomer
Styrene-butadiene/oil masterbatch	—	103.1	Elastomer–extender masterbatch
Cis-polybutadiene	—	25	Special purpose elastomer
Oil-soluble sulfonic acid	2.0	5.0	Processing aid
Stearic acid	2.5	2.0	Accelerator–activator
Zinc oxide	3.5	3.0	Accelerator–activator
Phenyl-β-naphthylamine	2.0	2.0	Antioxidant
Substituted N,N'-*p*-phenylene-diamine	4.0	4.0	Antiozonant
Microcrystalline wax	1.0	1.0	Processing aid and finish
Mixed process oil	5.0	7.0	Softener
HAF carbon black	50	—	Reinforcing filler
ISAF carbon black	—	65	Reinforcing filler
Sulfur	2.5	2.8	Vulcanizing agent
Substituted benzothiazole-2-sulfonamide	0.5	1.5	Accelerator
N-nitrosodiphenylamine	0.5	—	Retarder
Total weight	173.5	220.4	
Specific gravity	1.12	1.13	

[a] Parts per hundred parts of rubber, by weight.

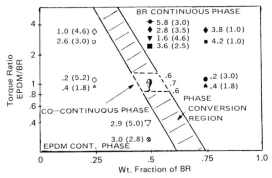

Figure 7.24. Morphology changes as function of Brabender torque ratio and composition. The numbers show the mean domain diameter in μm and the length-to-diameter ratio (in parentheses). $\triangledown$, E3/HV2; $\ominus$, E2/HV3; $\otimes$, E4/HV1; $\diamond$, E1/HV4; $\square$, E1/MV2; $\bigcirc$, E2/HV1; $\triangle$, E1/MV1. Shaded symbols indicate EPDM as the dispersed phase and open symbols, BR.[68]

natural rubber (IR), *cis*-polybutadiene (BR), butyl rubber (IIR), and ethylene propylene rubber (EPDM).

In most tire formulations, sulfur plus an accelerator forms the vulcanizing system. The rubber is reinforced with carbon black, and a number of other specialized ingredients are added, as shown in Table 7.8.[63] Although the art of rubber compounding is beyond the scope of this work, the subject has been widely reviewed.[60–63]

The art of rubber/rubber blending has also been reviewed recently.[60–67] One point of special interest, however, has to do with the development of dual-phase continuity. Avgeropoulos *et al.*[68] showed that a co-continuous phase structure of EPDM/BR blends form at equal weight fractions and equal viscosities (see Figure 7.24). Figure 7.24, of course, follows the general principles laid down in Figure 2.5 and concurrent discussion. The phase domain sizes attained by the blending, 0.2–4 μm, are comparable to solution polymerization techniques.

REFERENCES

1. Yu. S. Lipatov, L. M. Sergeeva, L. V. Mozzhikhina, and N. P. Apukhtina, *Vysokomol. Soedin.* **A16**(10), 2290 (1974).
2. Yu. S. Lipatov, L. M. Sergeeva, L. V. Karabanova, A. Ye Nesterov, and T. D. Ignatova, *Vysokomol. Soedin* **A18**(5), 1025 (1976).
3. Yu. S. Lipatov and L. M. Sergeeva, *Usp. Khim.* **45**, 138 (1976).
4. E. Helfand and Y. Tagami, *J. Chem. Phys.* **56**, 3592 (1972).
5. P. J. Flory and J. Rehner, *J. Chem. Phys.* **11**, 521 (1943).
6. D. L. Siegfried, D. A. Thomas, and L. H. Sperling, *Macromolecules* **12**, 586 (1979).

7. S. C. Hargest, J. A. Manson, and L. H. Sperling, *J. Appl. Polym. Sci.* **25**, 469 (1980).
8. J. L. Thiele and R. E. Cohen, *Polym. Prepr.* **19**(1), 137 (1978).
9. D. Klempner, H. L. Frisch, and K. C. Frisch, *J. Polym. Sci.* *A-2* **8**, 921 (1970).
10. J. A. Manson and L. H. Sperling, in *Polymer Blends and Composites*, Plenum, New York (1976), Chap. 3.
11. C. B. Bucknall, *Toughened Plastics*, Applied Science, Banking, Essex, England (1977).
12. A. A. Donatelli, L. H. Sperling, and D. A. Thomas, *Macromolecules* **9**, 676 (1976).
13. R. E. Touhsaent, D. A. Thomas, and L. H. Sperling, in *Toughness and Brittleness of Plastics*, R. D. Deanin and A. M. Crugnola, eds., American Chemical Society, Washington, D.C. (1976).
14. G. E. P. Box and K. B. Wilson, *J. R. Stat. Soc. Sect. B.* **13**, 1 (1951).
15. R. E. Touhsaent, D. A. Thomas, and L. H. Sperling, *J. Polym. Sci.* **46C**, 175 (1974).
16. P. R. Scarito and L. H. Sperling, *Polym. Eng. Sci.* **19**, 297 (1979).
17. S. C. Kim, D. Klempner, K. C. Frisch, and H. L. Frisch, *J. Appl. Polym. Sci.* **21**, 1289 (1977).
18. P. I. Donnelly, *Mechanical Properties of Polymers*, Wiley-Interscience, New York (1971), p. 259.
19. A. V. Tobolsky, *Properties and Structure of Polymers*, Wiley, New York (1960).
20. S. C. Kim, D. Klempner, K. C. Frisch, and H. L. Frisch, *Macromolecules* **10**, 1187 (1977).
21. G. P. Belonovskaya, J. D. Chernova, L. A. Korotneva, L. S. Andrianova, B. A. Dolgo-plosk, S. K. Zakharov, Yu. N. Lazanov, K. K. Kalninsh, L. M. Kaljuzhnaya, and M. F. Lebedeva, *Eur. Polym. J.* **12**, 817 (1976).
22. V. K. Belyakov, A. A. Berlin, I. I. Bukin, V. A. Orlov, and O. G. Tarakanov, *Polym. Sci. USSR* **10**, 700 (1968).
23. S. C. Kim, D. Klempner, K. C. Frisch, H. Ghiradella, and H. L. Frisch, *Polym. Eng. Sci.* **15**, 339 (1975).
24. N. Devia, L. H. Sperling, J. A. Manson, and A. Conde, *Macromolecules* **12**, 360 (1979).
25. N. Devia, L. H. Sperling, J. A. Manson, and A. Conde, *Polym. Eng. Sci.* **19**(12), 869 (1979).
26. N. Devia, L. H. Sperling, J. A. Manson, and A. Conde, *Polym. Eng. Sci.* **19**(12), 878 (1979).
27. N. Devia, L. H. Sperling, J. A. Manson, and A. Conde, *J. Appl. Polym. Sci.*, **24**, 569 (1979).
28. G. M. Yenwo, L. H. Sperling, J. Pulido, J. A. Manson, and A. Conde, *Polym. Eng. Sci.* **17**(4), 251 (1977).
29. H. L. Frisch, J. Cifaratti, R. Palma, R. Schwartz, R. Foreman, H. Yoon, D. Klempner, and K. C. Frisch, in *Polymer Alloys*, D. Klempner and K. C. Frisch, eds., Plenum, New York (1977).
30. S. C. Kim, D. Klempner, K. C. Frisch, and H. L. Frisch, *Macromolecules* **10**, 1191 (1977).
31. W. E. H. Davies, *J. Phys. D* **4**, 318 (1971).
32. J. C. Maxwell, *Treatise on Electricity and Magnetism*, 3rd ed., Vol. 1, Oxford University, London (1904).
33. C. J. F. Böttcher, *Theory of Electric Polarization*, Elsenier, Houston (1952).
34. B. Budiansky, *J. Mech. Phys. Solids* **13**, 223 (1965).
35. H. L. Frisch and K. C. Frisch, *Prog. Org. Coat.* **7**, 105 (1979).
36. J. W. Rees, U. S. Pat. 3,264,272 (1966).
37. J. W. Rees, U. S. Pat. 3,404,134 (1968).
38. L. Holliday, ed., *Ionic Polymers*, Wiley, New York (1975).
39. M. J. Lysaght, in *Polyelectrolytes*, K. C. Frisch, D. Klempner, and M. Patsis, eds., Technomic, Westport, Connecticut (1976).
40. G. S. Solt, Br. Pat. 728,508 (1955).

41. J. Seidl, J. Malinsky, K. Dusek, and W. Heitz, *Adv. Polym. Sci.* **5**, 113 (1967).
42. B. N. Kolarz, *J. Polym. Sci. Polym. Symp.* **47**, 197 (1974).
43. J. R. Millar, in *Kunstharz Ionentauscher*, Akademie, Berlin (1970), p. 43.
44. H. Czarczynska and W. Trochimczuk, *J. Polym. Sci. Polym. Symp.* **47**, 111 (1974).
45. W. Trochimczuk, in *Structure and Properties of Polymer Networks*, 9th Europhysics Conference on Macromolecular Physics, Jablonna, Poland, April 1979, European Physical Society, Warszawa, Poland.
46. B. N. Kolarz, Raport Seria PRE nr. 8, Politechniki Wroclawskiej, Wroclaw, Poland 1979.
47. W. Trochimuczuk, presented at the IUPAC sponsored conference, Modification of Polymers, Bratislava, Czechslovakia, July 1979.
48. J. H. Barrett and D. H. Clemens, U.S. Pat. 3,966,489 (1976).
49. M. J. Hatch, U.S. Pat. 3,957,698 (1976).
50. L. H. Sperling, V. A. Forlenza, and J. A. Manson, *J. Polym. Sci. Polym. Lett. Ed.* **13**, 713 (1975).
51. G. Lopatin and H. A. Newey, in *Reverse Osmosis Membrane Research*, H. K. Lonsdale and H. E. Podall, eds., Plenum, New York (1972).
52. U. Merten *Desalination* **1**, 297 (1966).
53. O. Kedem and A. Katchalsky, *Trans. Faraday Soc.* **59**, 1918, 1831, 1941 (1963).
54. S. C. Hargest, J. A. Manson, and L. H. Sperling, *J. Appl. Polym. Sci.* **25**, 469 (1980).
55. G. Akovali, K. Biliyar, and M. Shen, *J. Appl. Polym. Sci.* **20**, 2419 (1976).
56. G. C. Berry and M. Dror, *Am. Chem. Soc. Div. Org. Coat. Plast. Prepr.* **38**(1), 465 (1978).
57. M. Dror, M. Z. Elsabee, and G. C. Berry, *Biomater. Med. Devices Artif. Organs* **7**, 31 (1979).
58. L. H. Sperling and D. A. Thomas, U.S. Pat. 3,833,404 (1974).
59. P. Predecki, *J. Biomed. Mater. Res.* **8**, 487 (1974).
60. G. Kraus, ed., *Reinforcement of Elastomers*, Interscience, New York (1965).
61. H. J. Stern, *Rubber; Natural and Synthetic*, 2nd ed., Palmerton, New York (1967).
62. L. Bateman, ed., *The Chemistry and Physics of Rubber-like Substances*, Wiley, New York (1963).
63. H. L. Stephens, in *Rubber Technology*, 2nd ed., M. Morton, ed., Van Nostrand Reinhold, New York (1973).
64. E. T. McDonel, K. C. Baranwal, and J. C. Andries, in *Polymer Blends*, Vol. 2, D. R. Paul and S. Newman, eds., Academic, New York (1978).
65. P. J. Corish and B. D. W. Powell, *Rubber Chem. Technol.* **47**, 481 (1974).
66. B. N. Dinzburg, V. P. Popova, V. G. Dyunina, and A. E. Chalykh, *Koloidn. Zh.* **38**(2), 338 (1976).
67. D. F. Lohr, Jr., and W. Kanh, U.S. Pat. 3,928,282 (1976).
68. G. N. Avgeropoulos, F. C. Weissert, P. H. Biddison, and G. G. A. Böhm, *Rubber Chem. Technol.* **49**, 93 (1976).

ACTUAL OR PROPOSED APPLICATIONS

While the preceding chapter delineated the basic engineering and mechanical behavior of IPNs, this chapter will be devoted to selected applications, actual or proposed.* Many of these applications are discussed in the patent literature. Not so surprisingly, some of the more complex and/or intricate syntheses, as well as a wealth of detailed physical properties, are contained in these patents.

As per the definition of patentability, each patent addressed a novel method of synthesis, process, or application. Table 8.1[1] summarizes some 21 selected patents based on chemical crosslinking. Applications range from ion exchange resins, through adhesives, to impact-resistant plastics.

Newer on the scene are the physically crosslinked materials, which might be designated as the thermoplastic IPNs. In each of these cases, the materials can be melt processed but on cooling attain some of the characteristics of crosslinked materials. Many of these "blends" exhibit dual-phase continuity, as emphasized by Kresge.[2] There are several distinct ways of forming physical crosslinks, as indicated in Table 8.2. As discussed in Section 5.7, several independent ways of forming a thermoplastic IPN exist. Some of these have been explored, as will be illustrated below.

In addition, some patents describe the simultaneous blending and chemical crosslinking of multipolymer combinations. It is somewhat remarkable that conducting a crosslinking reaction during blending results in a material that has rheological properties suitable for the usual thermoplastic processing techniques. Kresge[2] gives an excellent review of the thermoplastic IPNs.

Strictly speaking, the use of the term "network" refers to the crosslinking of the single polymer chains. Therefore, an IPN implies a mutual solution of the two networks. While this problem was dealt with in Section 1.2, a brief review is instructive.

An IPN is defined as a combination of two (or more) polymers in network form, at least one of which is synthesized and/or crosslinked in the immediate presence of each other. The term "interpenetrating polymer

---

* This phraseology is inspired by the wide reference to patents. In only a very few cases has the literature identified a product with a description of an IPN.

Table 8.1. Selected Patents on IPNs and Related Materials[1]

No.	Author and patent No.	Polymer I	Polymer II	Use	Comments	Common designation
1	M. J. Hatch, U.S. Pat. 3,041,292 (1962)	Sulfonated polystyrene	Poly(acrylic acid)	Ion exchange resin	Suspension polymerization	Semi-I
2	T. A. Solak and J. T. Duke, U.S Pat. 3,426,102 (1969)	Polybutadiene	Poly(acrylonitrile-co-ethyl acrylate)	Impact-resistant plastic	Emulsion polymerization	Graft copolymer
3	R. D. Hibelink and G. H. Peters, U.S. Pat. 3,657,379 (1972)	Epoxy resin	Polyester resin	Adhesive	Suspension polymerization; polyester resin AB crosslinked, P_3 = polystyrene	IPN
4	F. G. Hutchinson, British Pat. 1,239,701 (1971)	Polyurethane	Poly(methyl methacrylate)	Shapable polymeric articles	Bulk polymerization	Semi-IPN
5	J. M. Hawkins, British Pat. 1,197,794 (1970)	Epoxy	Polyurethane	Adhesive	Bulk polymerization	Semi-IPN
6	G. S. Solt, British Pat. 728,508 (1955)	Chloromethyl polystyrene	Sulfonated polystyrene	Ion exchange resin	Suspension polymerization; complex reaction scheme	IPN
7	W. H. Parrkiss and R. Orr, British Pat. 786,102 (1957)	Polyester resin	Polystyrene	Flexible casting resin	P_3 = epoxy resin; bulk polymerization. Unsaturated polyester plus styrene produces AB crosslinked copolymer; two networks with three polymers	IPN
8	J. J. P. Staudinger and H. M. Hutchinson, U.S. Pat. 2,539,376 (1951)	Poly(methyl methacrylate)	Poly(ethyl vinyl ketone)	Optically smooth plastic surfaces	Bulk polymerization. Some compositions IPNs	Semi-I

—— continued overleaf

No.	Investigator, patent	Polymer I	Polymer II	Application	Comment	Type
9	J. J. Staudinger and H. M. Hutchinson, U.S. Pat. 2,539,377 (1951)	Polystyrene	Polystyrene	Optically smooth plastic surfaces	Bulk polymerization. Millar IPNs and Semi-Is. (Original application date in Great Britain July 23, 1941)	IPN
10	K. C. Frisch, H. L. Frisch, and D. Klempner, Ger. Pat. 2,153,987 (1972)	Polyacrylate	Polyurethane	Tough elastomeric films	Both polymers emulsion polymerized (Example 1)	IPN
		Epoxy	Polyurethane	Tough plastic	Simultaneous polymerization (Example 17)	SIN
		Polyester	Polystyrene	Tough plastic	P_3 = polyurethane (Example 12) (three distinguishable topologies in patent)	SIN
11	H. A. Clark, U.S. Pat. 3,527,842 (1970)	Polysiloxane	Polysiloxane	Pressure-sensitive adhesive	P_3 = polysiloxane. Active groups: P_1 has OH, P_2 has CH=CH$_2$, P_3 has H, P_2 and P_3 form an AB crosslinked copolymer	IPN
12	Anonymous (CIBA, Ltd.), British Pat. 1,223,338 (1971)	Poly(diallyl phthalate)	Epoxy	Compression molding composition	Simultaneous reactions	IPN
13	H. L. Stephens and T. F. Reed, U.S. Pat. 3,645,940 (1972)	SBR	Starch	Reinforced rubber	Mechanical mixing for phase inversion	Semi-IPN
14	L. H. Sperling and D. A. Thomas, U.S. Pat. 3,833,404 (1974)	Poly(ethyl methacrylate)	Poly(n-butyl acrylate)	Noise damping coating	Emulsion polymerization with overcoat	IPN
15	P. Mendoyanis, U.S. Pat. 3,316,324 (1967)	Polysulfide rubber	Epoxy resin	Adhesive sealant and crack filler	Components react on mixing in bulk	SIN or IPN

Table 8.1.—*Continued*

No.	Author and patent No.	Polymer I	Polymer II	Use	Comments	Common designation
16	B. Vollmert, U.S. Pat. 3,055,859 (1962)	Poly(butyl acrylate)	Polystyrene	Impact-resistant plastic	Grafting and crosslinking in bulk postpolymerization. (Example 1)	Semi-IPN
		Poly(butyl acrylate)	Poly(butyl acrylate-co-styrene)	Impact-resistant plastic	P_3 = polystyrene, P_4 = polystyrene; P_1 has carboxyl groups, P_2, P_3, and P_4 have OH, latex polymerization, mixed latex type (Example 7)	Semi-IPN
		Poly(n-butyl acrylate)	Poly(n-butyl acrylate-co-acrylonitrite)	Impact-resistant plastic	P_3 = poly(styrene-co-acrylonitrile). Latex polymerization; polymers mixed after precipitation (Example 20). This patent has 31 examples, mostly distinguishable topologies!	Two noninterpenetrating networks; a blend
17	J. H. Spiner, U.S. Pat. 3,681,475 (1972)	Butyl acrylate copolymer	Poly(butyl acrylate-co-tert-butyl-styrene)	Impact-resistant plastic	P_3 = poly(methyl methacrylate). Emulsion polymerization, no new particles (Example 7)	IPN
18	M. Baer, U.S. Pat. 3,041,309 (1962)	Acrylic copolymer	Methacrylic copolymer	Additive to plastics for toughening	P_3 = poly(styrene-co-acrylonitrile), emulsion polymerization	Graft + Blend
19	C. F. Ryan and R. J. Crochowski, U.S. Pat. 3,426,101 (1969)	Poly(n-butyl acrylate)	Polystyrene	Impact modifier	P_3 = poly(methyl methacrylate); P_4 = polyvinyl chloride). Latex polymerization, overcoat, blend	IPN + blend
20	J. Rosenberg, U.S. Pat. 3,928,113 (1975)	Water-soluble polymer	Polyene	Removable nail polish	Surface graft	Semi-II
21	D. F. Lohr, Jr., and J. K. Kang, U.S. Pat. 3,928,282 (1975)	Hydro-formylated polybutadiene	1,2 Poly-butadiene	Electrical insulation	Simultaneous grafting and crosslinking	IPN

Table 8.2. Types of Physical
Crosslinks Useful in IPN Formation

Block Copolymers
  Glassy blocks
  Crystalline blocks
  H-bonded blocks
Ionomeric sites
Semicrystalline homopolymer sites

network" was coined before the full significance of phase separation was realized. Since most of the more interesting IPNs are phase separated, the term "interpenetrating phases" may more accurately describe some systems. However, dual-phase continuity is not required in the definition of an IPN.

8.1. REINFORCED ELASTOMERS

This section and the one following will discuss the works of Hutchinson and co-workers,[3-8] who examined a wide range of IPNs, SINs, and semi-SINs. These patents illustrate the value of a polyurethane as polymer I, with either a vinyl polymer or an unsaturated polyester–styrene (ABCP) as polymer II. The more scientific aspects of some of the earlier patents are discussed in part in the series of papers by Allen et al.[9-14]

One of the basic concepts described by Hutchinson is called gel polymerization.[4] The general formula involves the preparation of a homogeneous composition comprising the precursers of a crosslinked polyurethane and a vinyl monomer. The urethane components are free of groups copolymerizable with vinyl monomer. While both polymerizations are simultaneous, the gelation of the polyurethane must be substantially completed before polymerization of the vinyl monomer is allowed to proceed to the extent that phase separation ensues. Then the conditions are changed so that polymerization of the vinyl monomer is completed.

If phase separation occurs before the polyurethane gelation, phase domains greater than 40 μm are common, i.e., gross phase separation occurs. However, proper control of the polymerization rates yields phases less than 2 μm in dimension, with the polyurethane forming the continuous phase and the vinyl polymer the discontinuous phase.

The preferred elastomers contain more than 50% by weight of polyurethane.[5] When the value of M_c of the vinyl polymer network is not greater than 2.5×10^3 g/mol, a decrease in the permanent set and a more linear stress–strain relationship was noted.

To illustrate the interrelationship between synthetic detail and mechanical behavior, Hutchinson (example 3 of Reference 5) prepared a solution containing 50 parts of a poly(ethylene adipate) having a molecular weight of 2000 g/mol and a hydroxyl value of 56 mg KOH g^{-1}, 0.7 parts of trimethylol propane, 8.6 parts of 4:4'-diphenyl methane diisocyanate, 0.1 part of t-butyl peroctoate, and 0.01 part of dibutyltin dilaurate to make up 80% of the total composition. The remaining 20% was composed of one of five different ratios of methyl methacrylate (MMA) and ethylene glycol dimethacrylate (EGDM) and contained 100 ppm Topanol stabilizer. After initial mixing at 60°C, the solution was degassed and charged to a mold. The mold was heated at 60°C for four hours to effect the polyurethane reaction. Thereafter, the mold and contents were heated at 80°C for 18 hr and 115°C for 2 hr to polymerize the methacrylate portion.

The properties of the final elastomeric materials are described in Table 8.3.[5] Several items bear emphasis: (a) The tensile strength goes through a maximum at midrange MMA/EGDM compositions. (b) Permanent set, a measure of the sample's "remembering" a deformation, decreases with increasing EGDM. (c) The modulus increases with increasing EGDM content, and the 100%, 200%, and 300% moduli progress in an orderly manner above 50/50 MMA/EGDM compositions.

8.2. SHEET MOLDING COMPOUNDS

Sheet molding compounds (SMC) have become increasingly important since their introduction into the plastics industry in the late 1950s. Nowadays, SMC is being used for a wide variety of transportation, appliance, and electrical applications.[15-17]

Table 8.3. Stress–Strain Studies on 80/20 Polyurethane/Poly(methyl methacrylate) SINs[5]

Proportion by weight MMA/EGDM	100:0	90:10	80:20	50:50	0:100
M_c in polymer of ethylenically unsaturated monomer	∞	1978	990	396	198
100% modulus, psi	480	550	700	860	1100
200% modulus, psi	440	510	650	900	1200
300% modulus, psi	600	660	1000	1200	1300
Elongation to break, %	450	450	370	360	340
Tensile strength, psi	4800	5700	5030	5300	5000
Permanent set, %	15	10	<5	<5	0
T_g, °C	−35	−35	−35	−35	−35

Conventional SMC is prepared from unsaturated polyester resin and styrene, together with magnesium or calcium oxides or hydroxides. The bases serve to thicken the styrene–polyester solution, probably through ionomer formation.[18,19]* Usually, SMC is used together with fillers such as calcium carbonate and glass fiber reinforcement for high modulus.

Ferrarini et al.[15] point out that the thickening reaction requires 1–5 days until the correct viscosity is reached. Sometimes the process exhibits variability from lot to lot, depending on particle size, degree of dispersion, etc. of the base.

As an improvement,[15] a crosslinked polyurethane can replace the alkaline earth induced thickness, and yield a superior product. They call the new process ITP,® for interpenetrating thickening process. The ITP process apparently combines the Hutchinson[3-8] technology with other advances.[20,21]

The urethane thickening reaction results in the formation of a polyurethane network in the presence of the unsaturated polyester resin, but substantially does not react with the polyester in many of the examples given. When the styrene-unsaturated polyester contained within the polyurethane network is polymerized, an interpenetrating polymer network is formed. Ferrarini, et al.[15] emphasize that the ITP gelled state is more than simply high viscosity, but gelled in the classic sense of the term. The urethane ingredients used are polyfunctional and react to give a crosslinked structure resulting in a network of chemical and physical bonds.

Apparently, the polyurethane reaction proceeds almost to the gel point before the final pouring and molding, then the reaction continues into the completed network stage during the styrene–polyester polymerization. Both the polyurethanes and the polyesters are very polar, and therefore a significant part of the gel network consists of physical bonds such as hydrogen bonds.[15] On heating, such bonds can break allowing proper casting, compression molding, or vacuum forming.

A significant part of the bonding may be developed by a cross reaction between the two networks, arising from a reaction between the isocyanates and the hydroxyls and/or the carboxyls on the unsaturated polyester.[8] This reaction, of course, leads to a grafted SIN.

Afterwards, the chemical bonds produce a thermoset and maintain a high degree of thermal stability in the molded parts. Details of the ITP resin, as used commercially, are also described elsewhere.[22]

Both the elastomeric polyurethanes[4,6] and plastic polyurethanes with high T_g were studied. Both yield reinforcement, albeit by different

---

* Since the styrene–polyester forms an AB crosslinked polymer, its ionomer are subjects of interest to this book in their own right (see Chapter 3).

mechanisms. While elastomeric polyurethanes increase impact strength and elongation, plastic-forming polyurethanes produce SINs with relatively high heat distortion temperatures and high flexural modulus.[8]

According to Hutchinson *et al.*[8] an optimum balance of properties will be obtained by preparing SINs from midrange compositions of near 50/50 polyurethane/polyester. The preferred polyisocyanate comprises at least one compound of the structure

$$OCN--X--NCO \qquad\qquad (8.1)$$

and the preferred structure is 4:4'-diphenylmethane diisocyanate.

Further, either the polyurethane or the polyester can be gel polymerized first.[8] In the first mode, the precursers of the crosslinked polyurethane may be substantially fully reacted together and with the ethylenically unsaturated polyester before any substantial reaction has taken place between the polyester and the vinyl monomer. Alternatively,[8] the unsaturated polyester may be substantially fully reacted with the vinyl monomer first. Both reactions may be run simultaneously.[8] Clearly, the morphology will be different in each case, as described in Chapter 6, i.e., the first polymer to reach gelation will tend to be the more continuous phase. Since the ITP process[15] utilizes the polyurethane reaction first, the behavior of the final SMC critically depends on the properties of the polyurethane.

Example 4 of Reference 8 shows the balance of properties obtained from a range of PU. The polyester solution (Crystic 199) was 38% by weight of styrene and 62% by weight of an unsaturated polyester. The polyurethane was based on 4:4'-diphenyl methyl diisocyanate and oxypropylated glycerol. The results are shown in Table 8.4.[8] The midrange compositions, samples B, C, and D, offer the best balance of properties and are clearly superior to the unmodified polyester, sample G. The polyurethane alone, F, it should be pointed out, has a T_g near 122°C.

In Example 21 of Reference 8, Hutchinson *et al.* describe the heat distortion characteristics of the materials (see Table 8.5.[8] Samples A and B were based on 4:4'-diphenyl methane diisocyanate, the latter prepared at a higher temperature. Sample C was prepared from a polyurethane composition having a T_g of −25°C. While the impact strength of C is much higher, it is a softer material with a lower heat distortion temperature. Since the material contains 60% polyurethane, the material is better described as a reinforced elastomer. Clearly experiment B yields the highest heat distortion temperature.

Table 8.4. Mechanical Behavior of Polyurethane/Polyester SINs Using Diphenyl Methane Diisocyanate[8]

Sample	PU/PE	Flexural modulus, kg cm^{-2}	Flexural breaking strength, kg cm^{-2}	Initial modulus, kg cm^{-2}	Tensile yield strength, kg cm^{-2}	Tensile breaking strength, kg cm^{-2}	Elongation to break, %	Impact strength, kg cm^{-2} Notched	Impact strength, kg cm^{-2} Unnotched
A	20/80	33,000	1150	—	none	—	—	—	9.1
B	40/60	34,000	1275	37,500	none	920	12	2.2	32.4
C	50/50	33,500	1150	35,000	800	700	13	2.1	34
D	60/40	31,500	935	33,000	740	690	14	4.2	87
E	80/20	28,000	870	29,000	760	—	—	4.9	116
F	100/0	22,000	890	27,000	510	440	13	5.4	>120
G[a]	0/100	33,500	820	39,000	—	430	18	—	6.7

[a] From example 2 of Reference 8 (see Column C of Table II in Reference 6).

Table 8.5. Mechanical Properties of 60/40 PU/PE SINs[8]

Experiment	Flexural modulus, kg cm^{-2}	Flexural yield strength, kg cm^{-2}	Impact strength kg cm^{-2} Notched	Impact strength kg cm^{-2} Unnotched	Falling weight impact strength, ft lb	Heat distortion temperature, °C 66 psi	Heat distortion temperature, °C 264 psi
A	33,300[a]	1390[a]	2.9	44.5	1.6	96	90
B	31,500[b]	1525[b]	6.0	45.0	3.6–4.2	120	114
C	2,500[b]	120[b]	9.9	7135	12–15	62	—

[a] Measured at a strain rate of 1% per minute.
[b] Measured at a strain rate of 10% per minute.

The Hutchinson *et al.* patent[8] also gives important examples relating to further shaping of a partially gel-polymerized article, Example 19, and fiber-reinforced systems, Examples 1 and 20. Clearly, the materials in actual commercial use as SMC will contain fibrous fillers. Hutchinson points out that, depending on the final properties desired, many different formulations are available.

In a somewhat related synthesis, Kircher and Pieper[23] describe a SIN composition of a polyurethane and poly(methyl methacrylate). They refer to their system as a one-pot process.

8.3. DYNAMICALLY PARTLY CURED THERMOPLASTIC ELASTOMER BLENDS

In a recent set of patents[24-30] Fischer disclosed a novel method of preparing thermoplastic elastomers from EPDM and isotactic poly-propylene, as well as other, chemically related species (EPM plus poly-ethylene, etc.) Fischer's work has led to the thermoplastic rubber known as TPR®.[31] A portion of this work, describing the blending operation, is described elsewhere.[32]

The basic feature involves blending polypropylene (10%–50%) with EPDM (50%–90%), plus less than the usual amounts of peroxide. The effect the dynamic semicure, the rubber, plastic, and curing agent are worked on an open roll mill, a Banbury mixer, or an extruder mixer.[24,25] During this time, between 60% and 93% of the material becomes insoluble in cyclo-hexane at 73°F. (Note: the plastic portion probably was already insoluble under these conditions; the reaction is partly gelling the elastomer.)

The EPDM/polypropylene blend was modeled by Kresge, to show the morphology of the material after crystallization.[2] He also prepared a binary blend of EPM copolymer and propylene. Lacking chemical crosslinking, the EPM could be extracted. A scanning electron micrograph is shown in Figure 8.1.[2] The white portions of the micrograph are polypropylene, and the dark voids are the areas where the EPM resided prior to solvent extraction. The polypropylene clearly exhibits phase continuity, resembling an open-celled sponge with micron-sized porosity. Careful examination of the micrograph indicates the voids also are continuous. Therefore, the material formed an interpenetrating two-phased structure. It is probable that the increase in viscosity of the EPDM phase, brought about by the partial crosslinking, effects the dual-phase character of the material.

In the case of the peroxide crosslinked blends, the dynamic aspects probably prevent the crosslinking phase from becoming truly continuous since the material will flow. The final blend is thus thermoplastic and can be

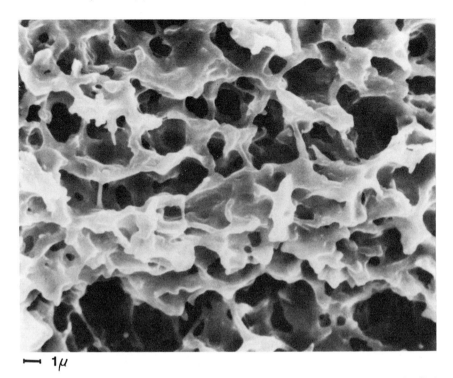

━━ 1μ

Figure 8.1. A model thermoplastic IPN. Electron scanning micrograph of a binary (70/30) blend of EPM and polypropylene prepared by intensive mixing after extraction of the EPM phase with heptane. The sample was fractured under liquid nitrogen; the bar is 1 μm.[2]

fabricated by such methods as molding and extrusion. Extruded insulation of wire, gaskets, flexible tubing, balls, weatherstripping, and flexible bumpers are mentioned as articles that can be made via this method. Some of the EPDM compositions may also crystallize slightly, improving the "thermoset" character of the material at use temperature.

Typical data are shown in Table 8.6 showing the effect of varying the peroxide level.[25] An 80/20 EPDM/polypropylene mix was processed for seven minutes at 360°F in each experiment. The peroxide employed was 2,5-bis(tert-butyl(peroxy))-2,5-dimethylhexane.

These materials, of course, are not strictly IPNs. However, the elastomer portion is partly cured and/or highly branched at the time of forming (somewhat similar to the urethane products discussed in Section 8.2). In addition, the polypropylene, being highly crystalline, contributes physical crosslinks to the system. The final material is composed of a partly gelled elastomer and a physically crosslinked plastic.

Table 8.6. Dynamically Partly Cured Thermoplastic Elastomer Blends[25]

Ingredients (parts)	Run No.			
	21	22	23	24
Peroxide	—	0.8	1.6	2.4
Gel, %	34	86	93	96
Swell, %	14	7	4	4
ML-5, 350°F	19	27	30	27
Tensile strength, psi	680	1040	1320	1210
100% Modulus, psi	670	720	790	780
Elongation, %	200	250	220	190
Elongation set at break, %	50	30	14	13
Hardness, shore A	73	77	78	78
Performance factor, psi × 10^3	2.7	8.7	21	17.7
Extrusion smoothness	10	10	9	0
Press moldable	yes	yes	yes	yes

As a thermoplastic elastomer, these materials behave somewhat similarly to the triblock copolymers developed by Holden and Milkovich[33] (see also Chapter 2). Above the melting point of the polypropylene plastic, the present material loses its thermoset characteristics. This compares to the glass transition softening of the polystyrene portion of the triblock copolymers.

It is of interest to review the molecular mechanism underlying the Fischer patents.

The facts are that the EPDM is the major phase and will crosslink with peroxide. The polypropylene is the minor phase and nonreactive under these circumstances. Also, the blend is formed first, and crosslinking starts during mixing.

It may be that the EPDM and polypropylene undergo a partial phase inversion during crosslinking. In any case, the increase in viscosity of the EPDM during crosslinking may cause it to become dispersed in the noncrosslinked polypropylene via a phase inversion.

Additionally, it is of interest to speculate on the mode of EPDM crosslinking during the mixing process which results in a thermoplastic material. One possibility involves the synthesis of sheet polymers, i.e., two-dimensional structures; strands, or very elongated spheres. As noted above, the crosslinking reaction takes place in a relatively high shear field. Molecular structures in a plane of constant shear would tend to remain in juxtaposition with each other, giving them time to react extensively. Portions above or below the plane in question, of course, constantly move relative to the plane, and have a reduced opportunity to react with the material in the plane. Thus, a compositon based on lamellae or cylinders, rather than a monolithic three-dimensional structure, might tend to be formed.

8.4. USE OF GRADED RUBBER LATEXES WITH SURFACE FUNCTIONALITY IN THERMOSETS

In 1974, Dickie and Newman[34-36] obtained three related patents on the synthesis of graded rubber latexes. The latexes contained surface functionality, and their use was in toughening thermoset plastics and coatings. The materials are applicable to the production of structural plastics such as automotive body panels, electrical appliance housings, boat construction, and conduits for the transmission of heated fluids.

The latexes themselves are basically semi-I IPNs or IPNs, prepared in a two-stage process. An acrylic, crosslinked seed latex serves as the core, while a plastic monomer mix II containing methacrylic or styrene monomers and monomers with reactive sites (epoxy, carboxy, hydroxy) is polymerized to form the shell of the latex. The graded composition is brought about by beginning to add the monomer II mix before the first monomer mix is completely polymerized.

This latex is dispersed in a thermoset prepolymer capable of reacting both with the reactive shell and a curing agent. The fully reacted thermoset may have three or four distinct polymers, variously crosslinked and/or grafted to form a series of complex IPNs. Besides their practical importance, the spatial topology of the several compositions is unique.

Example 1 of Reference 34 describes a four-polymer IPN. A seed latex of butyl acrylate and 1,3-butylene dimethacrylate forms C_{11}. Onto this seed latex core, a shell of methyl methacrylate and glycidyl methacrylate are polymerized, forming P_2. After coagulation, filtration, and washing, the resulting powder is dispersed in styrene and a stoichiometric amount of methacrylic acid. The dispersion is heated until the reaction between the methacrylic acid and the epoxy groups on the graded elastomeric particles is at least 50% complete.

The resulting dispersion is blended with an appropriate amount of an unsaturated polyester prepolymer prepared from orthophthalic acid, maleic anhydride, and neopentyl glycol, with a 5% excess of the last component, P_3. Molding polymerizes the styrene component, P_4, to form C_{34}. The overall composition is 20% rubber for a total of 40% by weight graded-rubber particles and 60% by weight of polyester–styrene present in weight ratio 65:35.

With slight simplification by omitting the latex graded structure, the structure may be written

$$C_{11}\,O_1\,P_2$$

$$O_C \qquad\qquad (8.2)$$

$$P_3\,O_C\,P_4$$

where the polystyrene chains, P_4, form AB crosslinked polymers simultaneously with the reactive shell of poly(methyl methacrylate), P_2, and the unsaturated polyester, P_3. The previously crosslinked poly(butyl acrylate), C_{11} is now interpenetrated by C_{24}. From a structural point of view, P_4 chains are simultaneously taking part in two distinct networks.

Some mechanical properties are shown in Table 8.7,[34] which describes the several stages required in improving the performance of polyester resins. When both graded rubber latex and chopped glass fibers are added to the polyester resin, the combination displays markedly superior strength, taken as a weighted average of elongation to break, stress at break, and modulus. In addition, Dickie and Newman[34] point out that this combination of reinforcing materials permits a much better retention of modulus on repeated stressing.

Example 5 of Reference 36 describes the preparation of two latexes, one reactive with the other. The graded rubber latex has a core of crosslinked poly(butyl acrylate) and a shell of linear poly(methyl methacrylate) containing hydroxypropylmethacrylate as the reactive species. The thermoset prepolymer emulsion is prepared from methyl methacrylate, containing glycidyl methacrylate as its reactive species. After polymerization, the two latexes are blended to yield a molding compound containing about 25% rubber, based on the core portions of the graded-

Table 8.7. Improvements in Mechanical Performance by Adding Rubber and Glass Fibers to Polyester Resins[34]

Composition	Elongation-to-Break, %	Stress at break	Modulus, psi
Polyester–styrene resin [a]	3.1	8,100	418,000
Graded rubber in resin[b]	7.5	5,700	292,000
Graded rubber in resin[c]	6.2	7,200	300,000
Polyester–styrene glass[d]	0.7	8,300	1,400,000
Polyester–styrene rubber–glass[d]	1.2	12,000	1,100,000

[a] Reference 34, Example 2, the plain resin.
[b] Reference 34, Example 1, described in text.
[c] Reference 34, Example 4, Latex shell is methyl methacrylate: styrene: glycidyl methacrylate, 35:30:30.
[d] Reference 30, Example 6. 20% of 1/8-in. chopped glass fibers added to product of Example 4.

rubber latex particles. The mixed emulsion is coagulated, washed, and a catalytic amount of 2-ethyl-4-methyl-imidazole is added. The final material is compression molded at 400°F for 15 min, during which time the glycidyl (epoxy) groups react with the hydroxy groups. (The shell of latex I reacts with latex II; both portions were linear PMMA.) Thus the final product has two networks: one composed of crosslinked poly(butyl acrylate), and the other composed of crosslinked poly(methyl methacrylate).

This material may also be profitably compared with the mixed latex material of Frisch et al.[37,38] In the latter case, the two latexes were blended and coagulated, but crosslinked intramolecularly.

8.5. DENTAL FILLINGS

Traditionally, silver amalgams and gold alloys have enjoyed a wide acceptance as dental restorative materials. While these metal alloys provided the desired toughness, ductility, strength, and ease of fabrication, they lacked the desired esthetics, and had higher coefficients of thermal conductivity than the natural tooth enamel.[39]

Around 1945, acrylic-type plastics were introduced as restorative materials.[39] These materials, prepared as a two-component package, were semi-II IPNs based on linear poly(methyl methacrylate) as polymer I, and MMA monomer plus crosslinking agents as network II.

Because of excessive shrinkage (about 5%), a high coefficient of thermal expansion, and poor abrasion resistance,[39] these materials were replaced by a series of composite materials based on a fundamental patent by Bowen.[40] The polymer portion was a single, densely crosslinked network formed by polymerizing the reaction product of glycidyl methacrylate with bisphenol A. The monomer, known as bis-GMA, had the structure

$$(8.3)$$

Important features in the Bowen patent[40] also included fumed silica treated with a vinyl silane coupling agent, resulting in a modern, reinforced composite material. This product was introduced by the 3M Company in 1964 as Addent. Subsequently, several improvements were made,[41–49] all resulting in single network materials.

Recently, interest has returned to the PMMA semi-II IPNs, however, but including inorganic fillers. In 1974, Masuhara et al.[50] described the use of trialkyl boron materials as initiators. They also improved the bonding strength of the composite to the tooth by pretreating the enamel with a silane. In their Example 5[50] they extend the tooth filling to orthopedic purposes for firmly joining bones. They employed poly(methyl methacrylate) as linear polymer I, and MMA plus ethylene glycol dimethacrylate to form network II. By using up to 85% inorganic fillers, wear resistance is improved, along with hardness and bonding. Undoubtedly, shrinkage on polymerization is also greatly reduced.

Other patents[51-53] also mention semi-II IPNs or graft copolymers (no crosslinker). The literature has been extensively reviewed.[39,54-56]

8.6. GRAFT COPOLYMERIZATION ONTO LEATHER

Although not widely recognized as such, leather is a crosslinked polymer based on collagen proteins. The materials commonly used for crosslinking collagen as part of the process of converting animal hides into leather fall into three main groups. Traditionally, vegetable tannins have been used for curing. More recently, aldehyde tannages and mineral tannages have become important. The most important mineral tanning agents are hydrated basic chromium III sulfate complexes.[57] The term "tanning," like the terms "curing" and "vulcanizing," was applied long before its chemical significance was understood. Each of these terms, of course, refers to a crosslinking process.

Animal skins are composed of collagen fibers arranged in a three-dimensional woven fabric (see Figure 8.2). The collagen itself is made up of three approximately equal length polypeptides arranged in the form of a triple helix, with a combined molecular weight of 300,000 g/mol. This triple helix is about 14 Å in diameter and 2900 Å long.[58] These molecules form into protofibrils which further aggregate to eventually form the fiber bundles depicted in Figure 8.2. Tanning makes the collagen nonputrescible, thus forming leather, one of the oldest known crosslinked polymers to be made and used.

Unlike many synthetic substitutes, leather lacks uniformity and is subject to chemical deterioration, water penetration, and abrasion damage, even though it is already a versatile material.

To improve the strength and elastic properties as well as fill in open areas, eliminate defects, and control water permeation, a series of vinyl, acrylic, and styrene type monomers were incorporated via an emulsion technique.[58-63] After polymerization, a semi-I IPN is formed if the vinyl monomer mix contain no crosslinker, and a full IPN is formed if a crosslinker is incorporated.

The general procedure involves adding water, initiator, redox system, and surfactant, to the skin in an appropriate container, and flushing with

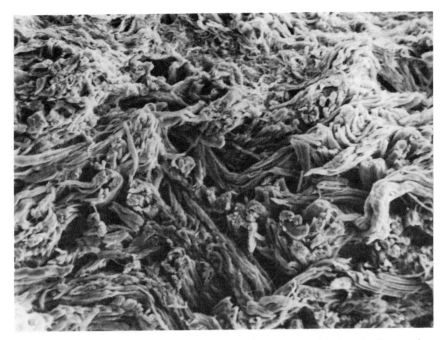

Figure 8.2. Scanning electron micrograph of sheepskin leather, showing the demarcation between the grain (top) and the corium (interior), 500 × magnification. Courtesy of F. P. Roper and R. J. Carroll, Eastern Regional Research Center, U.S.D.A., Philadelphia.

carbon dioxide. The monomer is added and the system tumbled. Typical details are shown in Table 8.8.[58] The monomers, including crosslinkers, are shown in Table 8.9.[58]

The properties of the grafted (IPN) leather are shown in Tables 8.10 and 8.11. Both crosslinking and ionic functional comonomers are illustrated, respectively. While the property changes are generally positive except for the tensile strength, the most significant change is the tensile elongation of the high-level (H) butyl acrylate compositions (Table 8.10). It may be that the 8% crosslinker employed was too high for maximum elongation; however, other properties come to the fore. For example, by choosing the appropriate combination of monomers, the grafted (IPN) leather can be made completely dry cleanable without loss of polymer.[64]

8.7. THERMOPLASTIC IPNs

The bulk of this monograph is naturally centered on materials prepared from two chemically crosslinked polymers. In Section 5.7, however, some

Table 8.8. Graft (IPN) Polymerization Method for
Chromium-Tanned Skin[58]

Step 1.	
Chrome-tanned skin[a]	100%
Water	800%
Emulsifier, Triton X-100	4%
Initiator	
Potassium persulfate	4%
Sodium bisulfite[b]	1.33%
Purge with CO_2, seal, and tumble for 30 min	
Step 2.	
Monomer	
for Low Level	40%
for High Level	100%
Purge with CO_2, seal, and tumble for 30 min	

[a] The skin is used in a wet drained state as received. All percentages are
based on the dry weight of skin being grafted.
[b] To avoid the release of SO_2, do not contact sodium bisulfite with acids.

of the scientific aspects of thermoplastic IPNs were already discussed.
These materials contain physical crosslinks rather than chemical crosslinks.
Examples include the block copolymers, ionomers, and crystalline polymers. Thermoplastic IPNs are formed when two such polymers are
chemically or physically blended.

Table 8.9. Monomers Used for IPN Formation in Leather[58]

Monomer	Abbreviation	Inhibitor, ppm[a]
n-Butyl acrylate	BA	MEHQ 5
Methyl methacrylate	MMA	MEHQ 10
1,3-Butylene dimethacrylate	BDMA	MEHQ 100
Ethylene diacrylate	EDA	HQ 50–140
Ethylene dimethacrylate	EDMA	MEHQ 50–100
1,6-Hexanediol diacrylate	HDDA	MEHQ 100
1,1,1-Trimethylol propane triacrylate	TMPTA	MEHQ 100
Dimethylaminoethyl methacrylate	DMAEMA	MEHQ 2000
Acrylic acid	AA	MEHQ 200
Methacrylic acid	MAA	MEHQ 100
Maleic acid	MalA	None
Fumaric acid	FumA	None

[a] MEHQ = Monomethyl ether of hydroquinone. HQ = Hydroquinone.

Table 8.10. Properties of Grafted Leathers: Semi-Is and IPNs[58]

Monomer system and level[a]		Percent polymer[b]		Percent change in [c]			
		Total	Bound	Thickness	Tensile elongation	Tensile load	Tensile strength
BA	H	44	16	43	160	70	9
BA–EDA	H	47	45	93	80	59	−19
BA–BDMA	H	55	48	159	216	38	−46
BA–HDDA	H	39	38	62	52	22	−25
BA–TMPTA	H	41	38	63	103	34	−18
MMA	H	50	16	70	−8	24	−30
MMA–EDMA	H	49	45	47	−21	28	−14
MMA–BDMA	H	59	50	140	38	64	−32
MMA–HDDA	H	38	35	59	14	20	−24
MMA–TMPTA	H	44	42	70	8	20	−28
BA	L	31	13	14	33	17	8
BA–EDA	L	24	23	42	52	31	−44
BA–BDMA	L	36	32	82	36	23	−30
BA–HDDA	L	24	23	32	29	24	−7
BA–TMPTA	L	25	25	16	19	19	4
MMA	L	30	13	32	6	23	0
MMA–EDMA	L	30	29	33	0	10	−16
MMA–BDMA	L	34	30	65	6	51	−10
MMA–HDDA	L	24	23	38	22	28	−6
MMA–TMPTA	L	24	23	28	2	18	−9

[a] The abbreviations used are taken from Table 8.9; H is high level, i.e., 100% of the dry substance of the leather; L is low or 40% of the dry substance of the leather. When used, the monomer to comonomer ratio is 23 to 2.

[b] The percent total polymer is

$$100 - \left(\frac{\text{percent N in grafted leather}}{\text{percent N in control leather}} \times 100 \right)$$

Percent bound polymer = percent total polymer − percent extractable polymer. (All values are on moisture and ash-free weights.)

[c] The percent changes are based on untreated control samples. A typical set of values for the unlubricated control pieces would be: thickness, 0.040 in.; elongation, 55%; load, 42 lb; and tensile strength, 2100 psi.

Blends of two crystalline polymers have been known for many years. Some of the early work centered on improving the low-temperature properties, particularly impact strength, of polypropylene by blending in greater or lesser amounts of polyethylene.[65–67] These early works can now be identified as crystalline/crystalline thermoplastic IPNs. These materials are the blend analog of the polyallomers,[68,69] which are multiblock copolymers of polypropylene and polyethylene.

Table 8.11. Properties of Semi-I IPN Leathers: Comonomers with Vinyl and Ionic Functionalities[58]

Monomer[a] system and level		Percent polymer[b]		Percent change in[c]			
		Total	Bound	Thickness	Tensile elongation	Tensile load	Tensile strength
BA	H	44	16	43	160	70	9
BA–AA	H	49	49	70	88	27	−30
BA–MAA	H	45	37	44	28	32	−5
BA–MalA	H	40	27	40	82	71	24
BA–DMAEMA	H	48	24	79	52	40	−22
MMA	H	50	16	70	−8	24	−30
MMA–MAA	H	54	49	88	−16	7	−37
MMA–MalA	H	49	26	82	42	79	−6
MMA–FumA	H	47	25	81	34	28	−26
MMA–DMAEMA	H	39	18	112	20	−8	−56
BA	L	31	13	14	33	17	8
BA–AA	L	29	27	34	58	41	5
BA–MAA	L	30	23	19	25	66	42
BA–DMAEMA	L	25	11	13	20	24	10
MMA	L	30	13	32	6	23	0
MMA–MAA	L	27	26	9	0	7	−6
MMA–DMAEMA	L	24	10	44	−4	22	−16

[a] The abbreviations used are taken from Table 8.9; H is high level, i.e., 100% of the dry substance of the leather; L is low or 40% of the dry substance of the leather. When used, the monomer to comonomer ratio is 23 to 2.
[b] The percent total polymer is

$$100 - \left(\frac{\text{percent N in grafted leather}}{\text{percent N in control leather}} \times 100 \right)$$

Percent bound polymer = percent total polymer − percent extractable polymer. (All values are on moisture- and ash-free weights.)
[c] The percent changes are based on untreated control samples. A typical set of values for the unlubricated control pieces would be: thickness, 0.040 in.; elongation, 55%; load, 42 lb; and tensile strength, 2100 psi.

More recently, blends of block copolymers and crystalline polymers have been developed by Davison and Gergen.[70,71] The block copolymers are based on hydrogenated styrene–butadiene–styrene triblock materials, known as styrene-ethylene butylene–styrene (SEBS) triblock copolymers. The crystalline polymers are polyamides[70] or thermoplastic saturated polyesters.[71]

Davison and Gergen[70] point out that their high-melting-point resins are normally incompatible with the SEBS block copolymers, and ordinary blends produce grossly heterogeneous mixtures with no useful properties. Their improvements arise from blending materials in such a way as to form interpenetrating phases, i.e. where one polymer would be thought of as filling the voids of the second polymer. While there is no molecular mixing, and separate and distinct phases are formed, they are not in a form that can lead to gross phase separation causing delamination. It should be emphasized that crosslinked interpenetrating phases are important in the formation of rubbery and leathery materials, or else the material is likely to behave as an adhesive rather than an elastomer.

Davison and Gergen continue with the requirements for forming a thermoplastic IPN[70]:

> Without wishing to be bound to any particular theory, it is considered that there are two general requirements for the formation of an interpenetrating network. First, there must be a primary network formed or in the process of forming in the shearing field. This requirement if fulfilled by employing the block copolymers of the instant invention having sufficiently high molecular weight to retain domain structure in processing. Second, the other polymer employed must be capable of some kind of chemical or kinetic reaction to form an infinite network from a disassociated melt. The polymer must possess sufficient fluidity to penetrate the interstices of the primary network. This second requirement is fulfilled by employing the instant polyamides.

Davison and Gergen[70] describe two methods (other than the absence of delamination) by which the presence of an interpenetrating network can be shown. In one method, a solvent dissolves away the block copolymer of a molded or extruded object, and the remaining polymer structure, comprising the plastic, retains the shape of the molded or extruded object. (See Figure 8.1 for the probable morphology.) They further[71] point out that the unextracted phase must be continuous because it is geometrically and mechanically intact. The extracted phase must have been continuous before extraction, since quantitative extraction of a dispersed phase from an insoluble matrix is highly unlikely. Emphasizing that these materials have "simultaneous continuous phases,"[71] they allude to electron microscopy studies.

The second method to determine interpenetration involves a mechanical property such as tensile modulus. The load is shared between both polymers, approaching the theoretical upper limit of the Takayanagi model for co-continuous versus parallel structures.

Gergen and Davison stress the importance of matching the viscosities of the two components (isoviscous mixing). This aspect must take into account both the temperature and the shear rate of the mixing process.

The thermoplastic IPNs have outstanding resistance to shrinkage and distortion on heat aging, resistance to ozone, and an improved balance of mechanical properties.

Illustrative Embodiment I in Reference 71 shows tough, leathery elastomers prepared from midrange compositions. Table 8.12 shows the basic blend composition, and Table 8.13[71] illustrates mechanical behavior.

Table 8.12. Composition of Co-Continuous Interlocking Network Phases[71]

Component	Composition by parts	
	A	B
25,000–100,000		
25,000 SEBS	100	100
Poly(butylene terephthalate)	100	70
Shellflex 790 extending oil	100	—
Tuffto 6050 oil	—	50
Polypropylene	10	10
Irganox 1010 antioxidant	0.2	0.2
Dilaurothiodipropionate antioxidant	0.5	0.5
TiO_2	5	5

As expected, composition B, containing less poly(butylene terephthalate), has a higher elongation at break than composition A. The presence of polypropylene in this and other compositions is noted.

These materials were made by mixing isoviscous components (3000–3200 poises at 260°C) by two passes through a Brabender.

A related material is described by de Candia et al.[72–74] They studied the physical and mechanical behavior of a polybutadiene/polar or ionic monomer mix, where the elastomer was crosslinked while simultaneously the monomer (methacrylic acid or magnesium methacrylate) was polymerized. Thus, a chemical/physical IPN was formed. Research by Shumskii, et al.[75] in low-pressure polyethylene blended with polyoxymethylene, and Nishi et al.[76] on poly(vinyl chloride)/polyester blends provides additional examples.

Table 8.13. Mechanical Behavior of Block Copolymer/Polyester Blends[71]

Property	Recipe No.	
	A	B
Hardness (shore A)	85	85
Tensile at break (psi) normal	900	1050
Parallel	700	1150
Elongation at break (%) normal	220	350
Parallel	140	160
100% Modulus (psi) normal	660	800
Parallel	880	1100

Chemically blended thermoplastic IPNs were made by Siegfried *et al.*[77] Compositions included the use of SEBS triblock copolymer as polymer I and poly(styrene-*co*-methacrylic acid) (10% MMA) as polymer II. These materials were made with a swelling process similar to that employed in the sequential IPNs. The carboxylic acid groups were neutralized with sodium hydroxide solutions while the materials were in a Brabender. Siegfried *et al.* also prepared the corresponding mechanically blended thermoplastic IPNs.

The properties of the chemical and mechanical blends are compared in Table 8.14.[77] The chemically blended material is shown to have a lower melt viscosity than the mechanically blended analog. Notably, elongation at break is higher, indicating a more elastic composition. The energy to break, estimated from the area under the stress–strain curve, is also higher for the chemically blended composition than for the mechanically blended material, although the modulus of the former is lower.

8.8. NOISE- AND VIBRATION-DAMPING MATERIALS

Semicompatible IPNs of low and high T_g polymers damp noise and vibrations over the intervening transition range. The motion of flexible chains over their stiffer neighbors may underlie this phenomenon.

The most common types of damping materials employed involve simple homopolymers or copolymers, with efficient damping limited to a temperature range of approximately 20–30°C, centered about the glass–rubber transition of the polymer involved.[78-94] Incompatible polymer blends and grafts,[95-97] with widely separated glass–rubber transition zones, exhibit two such damping ranges with little damping in between. Semicompatible polymer blends and grafts,[98,99] however, where the mixing of polymer I and

Table 8.14. Rheological and Mechanical Behavior of 50/50 SEBS/P(S-*co*-NaMAA) Thermoplastic IPN[77]

| | Blending method | |
Property	Chemical	Mechanical
Brabender Torque, m g, at 180°C	1600	2050
Tensile strength, psi	3.7×10^3	2.4×10^3
Percent strain at break	410	200
Young's modulus, psi	3.1×10^4	4.5×10^4

polymer II is extensive but incomplete, lead to a broader temperature use range.

One way of attaining the required intimate mixing is through the use of IPNs.[100–102] If the free energy of mixing of the two polymers is near zero, phase domains of the order of 100 Å develop (see Figure 6.2). For ease of application as coatings, IPNs in latex form rather than bulk form were investigated.[100–102]

8.8.1. Theory of Broad-Temperature Damping

When polymers are in their glass transition region, the time required to complete an average coordinated movement of the chain segments approximates the length of time for the measurement. If dynamic or cyclical mechanical motions are involved, the time required to complete one cycle, or its inverse, the frequency, becomes the time unit of interest. At the glass transition conditions, which involve both temperature and frequency effects, the conversion or degradation of mechanical energy to heat reaches its maximum value. The degradation of vibrational energy to heat occurs when a polymer at its glass transition temperature is in contact with a vibrating surface.

The theoretical aspects of resonant vibration attenuation by coatings have been described by Ungar,[90] who described two main types of coating configurations: extensional and constrained. An extensional damping treatment is a single-layer coating in which energy dissipation (and consequent damping) evolves primarily from the flexural and extensional motions of the damping layer. A constrained layer treatment, Figure 8.3, consists of a two-layer system with a viscoelastic layer under a stiff constraining layer. The addition of the constraining layer produces a shearing action within the viscoelastic layer as the composite panel vibrates. The shear action in combination with flexure and extension greatly increases the amount of energy dissipated per cycle over extensional systems.

The damping terms follow from basic viscoelastic behavior (see Chapter 2). Extensional damping follows the loss modulus, E''. On the other

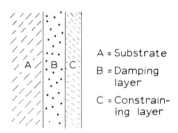

A = Substrate

B = Damping layer

C = Constraining layer

Figure 8.3. Constrained layer damping configuration.[102]

hand, constrained layer damping follows the loss tangent, tan δ, which equals E''/E'. The maxima in E'' and tan δ occur at somewhat different temperatures. The third term is the percent critical damping.

By measuring the vibrational decay rate, the logarithmic decrement, Δ, may be determined. Thus,

$$\Delta = -\frac{1}{N} \ln \frac{\chi_N}{\chi_0} \tag{8.4}$$

where N equals the number of cycles, and χ_0 and χ_N equal the initial amplitude, and the amplitude after N cycles, respectively. The percent critical damping, %C.D. is given by

$$\%\text{C.D.} = 100(2\pi\Delta) \tag{8.5}$$

(At 100% critical damping, a vibrating reed just barely returns to its initial configuration after one cycle.)

Examples of the waveforms generated by a highly damped structure and an undamped structure are given in Figure 8.4.

8.8.2. Constrained Layer Damping with IPNs

Methacrylic/acrylic latex IPNs, Section 5.4.1, yield convenient semi-compatible compositions. Tables 8.15 and 8.16[101] delineate several experimental and commercial compositions. Poly(vinyl acetate) copolymers, appropriately plasticized and filled, are the most important group of the latter materials.

Figure 8.5[101] compares formula A latex IPN constrained with epoxy resin, with various commercial materials over a broad temperature spectrum. Whereas the commercial materials, B, C, D, E, and F, all tend to peak sharply at one temperature, A shows relatively high damping throughout the temperature range.

Figure 8.6[101] demonstrates the damping capabilities of formula G in both extensional and constrained layer modes as a function of temperature. In the latter mode, a constraining layer material based on reinforced epoxy resin was used.

The advantages of the latex IPN constrained layer damping system are twofold:

(1) Materials can be damped over a broad and controllable temperature range.

(2) As a coating, both the damping and constraining layers can be applied over arbitrarily curved surfaces.

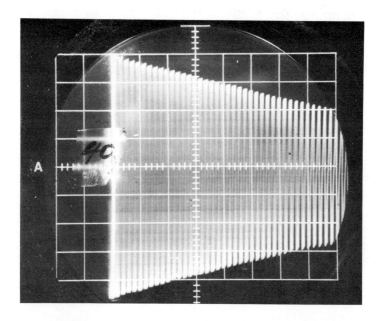

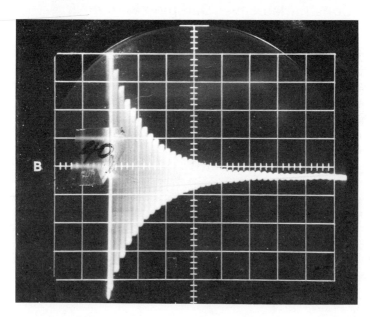

Figure 8.4. Oscilloscope traces of vibration decay: (A) undamped reed; (B) highly damped reed.[103] (The label 40 on the scope is the group number of the experiment.)

Table 8.15. Composition of Damping Polymers[94]

Designation	Source	Composition
A	IPN	75/25 poly(ethyl methacrylate)/(*n*-butyl acrylate)
B	commercial	poly(vinyl acetate)
C	commercial	poly(vinyl acetate)
D	commercial	unknown
E	commercial	unknown
F	commercial	unknown
G	IPN	(9-*co*-21)/(21-*co*-49) poly(ethyl methacrylate-*co*-ethyl acrylate)/(*n*-butyl acrylate-*co*-ethyl acrylate)
H	random copolymer	55/45 ethyl acrylate/methyl methacrylate
J	random copolymer	70/30 ethyl acrylate/methyl methacrylate
K	IPN	50/(30-*co*-16) poly(vinyl chloride)/poly(butadiene-*co*-acrylonitrile)

Table 8.16. Composition of Constraining Layers[94]

Designation	Composition
M	HS7130 reinforced with 5% Fybex
N	15 mil aluminum
P	Epon 828
Q	Epon 828 reinforced with fiber glass mat
R	Epon 828 reinforced with fiber glass cloth
S	13.5 mil steel
T	Laminate of Kevlar fabric impregnated with phenolic

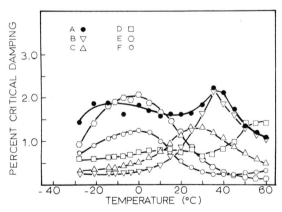

Figure 8.5. Temperature dependence of damping for formula A and commercial materials.[103]

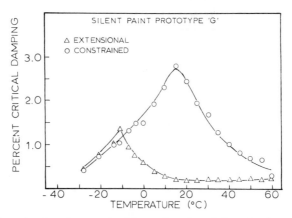

Figure 8.6. Extensional and constrained layer damping of formula G as a function of temperature.[103]

8.9. COATINGS AND ADHESIVES

The use of coatings and adhesives dates back to the beginning of civilization. Certainly Noah would not have stayed afloat for 40 days without pitch (Genesis 6:14). Once equipped with fire, pots, and stirrers, early humans began mixing and cooking many organic combinations. The sap or pitch of many plant species yielded their rubber or resinous components; boiled-down fish broths, horns, or other animal parts yielded protein animal glues; starchy or fibrous plant products, now identifiable as polysaccharides grew in abundance; and a host of other materials including eggs, bitumen, cheese, etc. were available. Whether any of these materials were IPNs is difficult to say, but many of these products are crosslinked chemically or physically in the use state. The early history of adhesives has been reviewed by Skeist (Chapter 1 in Reference 103).

8.9.1. Modified Phenol–Formaldehyde Resins

The first completely synthetic resin, based on phenol and formaldehyde, was invented by Baekeland and others in the first decade of this century.[104] While the phenol–formaldehyde resins yielded excellent adhesives then as now, blends with carboxylic rubber, epoxy resins, neoprene, nitrile rubber, and other polymers improved adhesive properties for various purposes.[103]

The phenol–formaldehyde resins are formed by a condensation reaction between the two components, made by splitting out water. In general, any phenol derivative and any aldehyde may be used. An incompletely reacted product, still soluble, of para-butyl phenol and formaldehyde is

$$ \text{(8.6)} $$

It must be emphasized that since the phenolics are crosslinked in final form, and most of the added polymers are also crosslinked in the final product, in effect these materials form IPNs, with greater or lesser extents of grafting between the two networks. It seems that phenol–formaldehyde resins were already being modified with other polymeric materials even as the original products were being developed. In 1910, Berend[105] described the preparation of a phenol–formaldehyde resin modified with animal glue. The mixture contained glue, phenol, glycerol, and formaldehyde.

Baekeland[106] himself mixed natural rubber with a phenol–formaldehyde resin at an intermediate stage of condensation. Catalysts made the resin infusible on vulcanization of the rubber. Other very early products of rubber blended with phenol and formaldehyde or other aldehydes were described by McGavack,[107–108] Aylsworth,[109] Frood,[110] and exhaustively reviewed by Ellis (Reference 111, Chapter 20).

In somewhat more recent work, it was shown that phenolic resins may be blended with nitrile rubber to bond a wide variety of construction materials, including wood, iron, glass, cardboard, and a number of synthetic polymers (see Chapter 17 in Reference 103, and References 112–114).

More recently, Tawney et al.[115,116] studied the vulcanization of butyl rubber with phenol–formaldehyde derivatives. These workers pointed out that during vulcanization of natural or butyl rubber, two competing reactions occur. One is the looked-for crosslinking reaction, and the other is reversion or devulcanization. This problem was overcome in the presence of phenol–formaldehyde components by the use of 2,6-dimethylol-4-hydrocarbyl-phenols as vulcanizing compounds. The butyl rubber/phenolic IPN exhibits improved high-temperature properties over ordinary butyl rubber.

A modified phenol–aldehyde adhesive system was described by Armstrong Cork Co.[117] The curable rubber-resin SIN was based on two polymer/prepolymer components that were mechanically mixed:

    1. A rubbery butadiene–acrylonitrile copolymer, cured with sulfur plus accelerator.

    2. A phenol–aldehyde resin containing reactive phenolic hydroxyl groups, plus an epoxide resin. After heat curing, high degrees of tensile, shear, and peel strengths were obtained.

8.9.2. Modified Alkyd Resins

    Another early material, still in wide use, that formed polymer networks, was the alkyd resins. The alkyds are a group of resins which can be called oil-modified polyesters. These polymers are formed through the condensation and exchange reactions among polybasic acids, polyhydric alcohols, and fatty monobasic acids. Typical polybasic acids are phthalic acid anhydride, maleic anhydride, etc. The polyhydric alcohol may be glycerin, pentaerythritol, etc., while the fatty acids may originate in soy bean or linseed oil. A typical alkyd has the following structure:

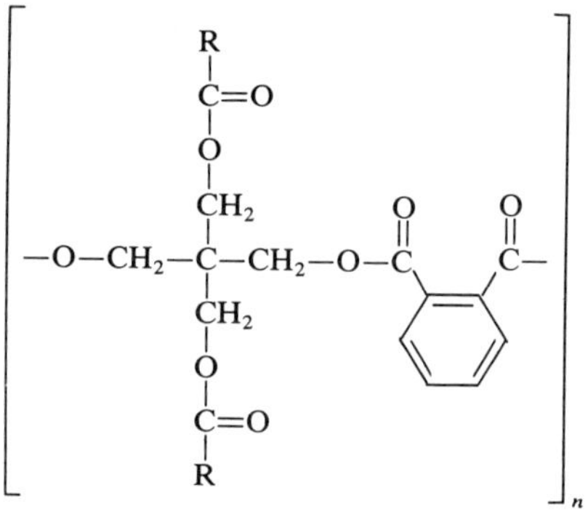

$$(8.7)$$

where R = fatty acid chain. In the fully reacted state, they are crosslinked through the polyhydric alcohol moiety. These, too, were blended with other polymers to make early versions of IPNs (Reference III, Chapter 42). For example, a resin based on phenyl glycidyl ether and anhydride was blended with rubber, and the mixture vulcanized to yield a rubbery composition of increased chemical and electrical resistance.[118] From an early date, great interest was expressed in compositions of the ubiquitous polystyrene in alkyd resins to form semi-II IPNs.[119] Some of the materials that were blended with alkyd resins, see Table 8.17, were reviewed by Ellis (Reference 111, Chapter 45).

Table 8.17. Blends or IPNs of Various Polymers with
Alkyd Resins[a]

Materials	Methods of blending with alkyd resins
Mastic	a,b,c,d
Dammar	a,b,c,d
Rosin	a,b,c,d
Shellac	c,d
Copal	a,c
Rubber (smoked sheet)	a,c,e
Water glass	f
Glue	c,f
Pitches	c,e
Casein	c,f
Chical	c,d
Rubber latex	c,f
Nitrocellulose	c,d
Pine tar	a,b,c,d
Canada balsam	a,b,c,d
Drying oils	a,b,c,d,f
Phenolic type of resin	a,c,d
Urea resins	a,c,f
Vinyl acetate	c,d
Paracumarone resins	a,b,c,d

[a] Reference 111, Chapter 45.
a. Cooking with other ingredients in preparation of alkyd resin complex.
b. Simple heat-blending.
c. Hot mixing on rollers, in dough mixers, or in Banbury mixers.
d. Dissolving both in a common solvent or miscible solvent.
e. Dissolving both in a common solvent or miscible solvent provided that the resin is of hydrocarbon-soluble type.
f. Using an aqueous ammonia or triethanolamine solution.

More recently, a number of interstitially crosslinked coating materials have been prepared.[120] These are blends in which a water-soluble or water-dispersible resin, capable of crosslinking, is blended with an ordinary thermoplastic emulsion. The main resins include aminoplast, phenol–formaldehyde, polyurethane, epoxy, and drying oils. Applications include sealants, adhesives, and architectural coatings.

Solomon[121] recently reviewed the subject of vinyl and acrylic modified oils and alkyds. While his emphasis is on the grafting reactions, of course all of these materials are semi-IPNs, since the oil or alkyd crosslinks during polymerization.

8.10. MISCELLANEOUS IPN PATENTS AND STUDIES

The previous sections have considered selected groups of patents, which illustrated in some depth particular ways of combining two or more polymers to create an IPN or related material. By their very definition, each patent addresses a different invention. Briefly, other concepts will now be introduced.

8.10.1. Crosslinked Acrylic Elastomer Latexes

A number of patents have emphasized acrylic elastomers as a method of rubber-toughening plastics.[122–124] The latexes are crosslinked, in part, to impart mechanical stability during processing. Acrylic latexes have improved stability to light and heat over SBR or NBR, and thus offer an improvement to ABS materials. In general, saturated elastomers have proved superior to the diene types for outdoor use because of their superior weather resistance.

Ryan and Crochowski[125] teach the use of acrylic latex IPNs dispersed in PVC copolymers to produce transparent, impact-resistant vinyls (see Table 8.1). A three-stage polymerization of the latex particles is required. A crosslinked rubbery latex such as poly(butyl acrylate) makes up the seed latex. A crosslinked vinyl aromatic, such as polystyrene, makes up network II. A linear poly(alkyl methacrylate), such as PMMA, forms polymer III. The finished latex is coagulated and blended with PVC to produce a tough, transparent plastic. Transparency is achieved by a close match of refractive indices.

Spilner,[126] in Example 7, prepared a two-stage latex IPN, based on butyl acrylate copolymers. The latex is dispersed in an MMA monomer mix (Example 13) to produce tough acrylic plastics after polymerization. Materials of this nature also exhibit a high degree of clarity.

Vollmert[127] presented 31 examples of multipolymer compositions, some of which were IPNs, (see Table 8.1). In example 20, a crosslinked poly(n-butyl acrylate) makes up network I. Poly(n-butyl acrylate-co-acrylonitrile) crosslinked with 1,4-butane-diol diacrylate makes up network II on a separate latex. A linear poly(styrene-co-acrylonitrile) latex makes up polymer III for a third latex. The three latexes are blended to form an impact-resistant polystyrene. This particular product, however, is not an IPN, because the two crosslinked latexes were polymerized and crosslinked separately and then mechanically blended together.

Johnson and Labana[128] prepared latex-based, acrylic IPNs by synthesizing a crosslinked seed latex, overcoated with an —OH bearing linear polymer II, which was subsequently crosslinked after molding. In their Example 1, a seed latex of butyl acrylate crosslinked with 1,3-butylene dimethacrylate was formed. Linear polymer II consisted of poly(methyl methacrylate-co-hydroxyethyl methacrylate), which formed polymer II of the core–shell latex structure. After coagulation and dispersion in acetone, a blocked diisocyanate was added. On compression molding, finally, the diisocyanate reacted with the hydroxyl groups to crosslink the PMMA and make an IPN.

Recently, latexes were prepared with a crosslinked core, crosslinked interlayers, and a crosslinked shell. A seed latex was overcoated five times with various combinations of methyl methacrylate and butyl acrylate, allyl methacrylate serving as the crosslinker.[129] Moldable products with good transparency, stress-clouding resistance, as well as impact and weathering resistance were attained. The subject was recently reviewed.[13]

8.10.2. MBS Resins

The term "MBS" refers to methyl methacrylate–butadiene–styrene resins, where methyl methacrylate has replaced the acrylonitrile in ABS. When dispersed in poly(vinyl chloride), a refractive index match can be obtained, which combines high clarity with impact resistance.

Like ABS resins, MBS can be made in a variety of ways. Amagi et al.[131] prepared a seed latex of SBR crosslinked with triethylene glycol dimethacrylate (Example 1).[131] A styrene–methyl methacrylate copolymer, also crosslinked with TEGDM, formed the second layer. Network III was poly(methyl methacrylate), again crosslinked with TEGDM. This forms a triple latex IPN. The material was salted out, and blended with PVC at a level of 15% to produce a Charpy impact strength of 85 kg cm/cm^2, combined with high clarity. Stress-whitening on bending was also reduced.

8.10.3. Unsaturated Polyester–Styrene Compositions

An unsaturated polyester dissolved in polymerizing styrene forms an AB crosslinked polymer network. When combined with another thermosetting resin, an IPN, or more usually, an SIN, is formed.

Useful for solid dielectric and protective coatings for electrical apparatus is a polyester–styrene/epoxy–chlorinated anhydride composition.[133] The chlorinated anhydride is prepared by reacting chlorendic anhydride (hexachloroendomethylenetetrahydrophthalic anhydride) with glycerol. The chlorine groups impart flame resistance.

Unsaturated polyester–styrene compositions were mixed with castor oil and TDI to form a polyurethane network[133] (Example 6). Such compositions make electrical ballasts with excellent heat dissipation and also reduce noise. As described in Section 6.4.2.2, castor oil–urethanes make excellent elastomers. The SIN with polyester–styrene plastic undoubtedly makes a sufficiently compatible mixture to impart noise damping, as described in Section 8.8.

Novel adhesives and casting resins have been made by the use of unsaturated polyester–styrene/epoxy resins,[134–139] imparting flexibility, improved low-temperature properties,[136] and underwater use.[135] Example 1 of Reference 134 mentions the use of the glycidyl polyether of castor oil as part of the epoxy formulation.

Arc resistance and electrical insulation, combined with low volatility, are mentioned[137] for polyester–styrene polymers containing cured aminoplasts as fillers. Curable epoxy resins may replace the polyester–styrene compositions.[137] Reinforced unsaturated polyester–styrene compositions with PVC[138] or other thermoplastic vinyl halide form a semi-II IPN. A separate multivinyl crosslinker is added to the system.[138] Unsaturated polyester–styrene/epoxy resins containing bentonite–amine reaction products materials that can be cured to a solid state without undergoing substantial volume shrinkage[139] are also tough and flexible.

8.10.4. Epoxy-Based Compositions

Semi-SINs using amine-cured epoxies and butadiene–acrylonitrile rubber[140] were found to yield adhesives with extremely high bonding strength to metal. The rubber was based on Hycar, and the epoxy was based on a range of compositions, including epoxidized soy bean oil (Example 2). Sometimes the rubber was sulfur cured, yielding an IPN.

A curable epoxy resin containing a crosslinked elastomeric latex[141] yields a special type of sequential IPN. The elastomers mentioned include SBR, EPR, NBR, silicone rubber, and PEA. Hawkins[142] described an epoxy/polyurethane semi-SIN, and Mendoyanis[143] revealed an epoxy/liquid rubber SIN. The liquid rubbers were based on polysulfide rubber or polyethylene. The final products could be extended to 400% at 0°F.[143]

Prevorsek and Aharoni[144] recently prepared a copolyester/polyepoxide semi-II IPN, which possessed a high green strength and tack before curing, and exhibited a single glass transition temperature. The polyepoxide was a mixture composed of high, intermediate, and low molecular weight components.

8.10.5. Polyurethane-Based Compositions

One of the more important patents in IPNs, by Frisch et al.[145] reveals a series of IENs (see Section 5.3) and SINs (see Section 5.5 and Table 8.1).

Each of the Frisch compositions discloses a polyurethane as one of the networks.

A composition composed of an isocyanate-terminated prepolymer, a multifunctional acrylic such as TEGDM or TMPTM, and an isocyanate-terminated polyhydric polyalkene ether on curing, forms a complex, grafted SIN.[146]

8.10.6. Silicone-Rubber-Based Compositions

An hydrophilic/hydrophobic SIN composition suitable for soft contact lenses was disclosed by Falcella et al.[147] Hydroxy ethyl methacrylate or similar monomers formed network I, and a polysiloxane formed network II.

Clark[148] disclosed an adhesive composition based on three PDMS linear polymer derivatives. Polymer I had —OH groups, attached to silica atoms, polymer II had —CH=CH$_2$, and polymer III had —H groups. After application, the Si—H groups reacted with the Si—CH=CH$_2$ to form network I. On further heating, polymer I self-crosslinks to make a type of SIN. Thus, good bonding and tack can be combined at the intermediate stage.

8.10.7. Natural-Rubber-Based Compositions

The use of natural rubber latex to form graft copolymers (and eventually IPN-related materials) was one of the earlier methods of preparing rubber/plastic compositions.[149–154] These materials are known as Heveaplus, which is a generic name of a series of raw materials made by graft copolymerization with other polymeric or resinous substances.[150] Heveaplus MG, prepared by polymerizing methyl methacrylate in the rubber latex, has attracted the most attention.

The latex itself retains a degree of thermoplasticity but can be vulcanized after molding[150] to make a semi-IPN. Undoubtedly, the MMA polymerization also introduces some ABCP copolymer formation.

8.10.8. Low-Profile Polyesters

Unsaturated polyesters, dissolved in greater or lesser quantities of styrene monomer, form the matrix material widely used for glass-fiber-reinforced thermoset sheet molding and bulk molding compounds. Upon polymerization of the styrene and concomitant conterminous grafting, an AB crosslinked copolymer network forms (see Section 5.6).

These materials suffer poor surface appearance, fiber pattern showthrough, etc., which arises during the 7% or so shrinkage during polymerization.[155,158] The surface can be smoothed, or made low-profile, by the addition of polymeric adjuvants.

Until recently, thermoplastic polymers have been employed as the adjuvants. The added polymers have included poly(vinyl acetate),[156] or

Table 8.18. Actual or Anticipated Uses for IPN and Related Materials

References[a]	Mode of combination	Application
a	Natural leather/rubber	Improved leather
b	Anionic/cationic	Piezodialysis membranes
c	Anionic/cationic	Ion-exchange resin
d	Plastic/rubber	Noise damping
e	Rubber/plastic semi-IPN	Impact resistant plastic
f	Plastic/plastic	Optically smooth surfaces
g	Rubber/rubber	Pressure-sensitive adhesive
h	Plastic/plastic	Compression molding composition
i	Plastic/rubber	Tough plastic
j	Water swellable/water swellable	Soft contact lenses
k	Rubber/plastic	Impact modifier
l	Rubber/plastic SINs	Adhesive coatings
m	Rubber/water swellable	Arteriovenous shunts
n	Plastic/plastic	Denture base materials
o	Plastic/plastic semi-II	Dental fillings
p	Plastic/plastic SINs	Sheet molding compounds
q	Rubber/crystalline plastic	Thermoplastic elastomers
r	Block copolymer/crystalline plastic	High-temperature elastomers
s	Unsaturated polyester–styrene/ crosslinked suspension	Low-profile plastics

[a] References:
  a. S. H. Feairheller, A. H. Korn, E. H. Harris, E. M. Filachione, and M. M. Taylor, U.S. Pat. 3,843,320 (1974).
  b. L. H. Sperling, V. A. Forlenza, and J. A. Manson, *J. Polym. Sci. Polym. Lett. Ed.* **13**, 713 (1975).
  c. G. S. Solt, Br. Pat. 728,508 (1955).
  d. L. H. Sperling and D. A. Thomas, U.S. Pat. 3,833,404 (1974).
  e. B. Vollmert, U.S. Pat. 3,055,859 (1962).
  f. J. J. P. Staudinger and H. M. Hutchinson, U.S. Pat. 2,539,377 (1951).
  g. H. A. Clark, U.S. Pat. 3,527,842 (1970).
  h. Anonymous (Ciba, Ltd.) Br. Pat. 1,223,338 (1971).
  i. K. C. Frisch, H. L. Frisch, and D. Klempner, Ger. Pat. 2,153,987 (1972).
  j. J. J. Falcetta, G. D. Friends, and G. C. C. Niu, Ger. Offen. Pat. 2,518,904 (1975).
  k. C. F. Ryan and R. J. Crochowski, U.S. Pat. 3,426,101 (1969).
  l. H. L. Frisch, J. Cifaratti, R. Palma, R. Schwartz, R. Forman, H. Yoon, D. Klempner, and K. C. Frisch, in *Polymer Alloys*, D. Klempner and K. C. Frisch, eds., Plenum, New York (1977).
  m. P. Predecki, *J. Biomed. Mater. Res.* **8**, 487 (1974).
  n. B. E. Causton, *J. Dent. Res.* **53**(3), 1074 (1974).
  o. L. N. Johnson, *J. Biomed. Mater. Res. Symp.* **1**, 207 (1971).
  p. F. G. Hutchinson, R. G. C. Henbest, and M. K. Leggett, U.S. Pat. 4,062,826 (1977).
  q. W. K. Fischer, U.S. Pat. 3,806,558 (1974).
  r. W. P. Gergen and S. Davison, U.S. Pat. 4,101,605 (1978).
  s. D. R. Stevenson, U.S. Pat. 4,048,257 (1977).

poly(methyl methacrylate), polystyrene, or saturated polyesters,[155] among others. After curing, these materials are all semi-IPNs. While the exact mechanism of shrinkage prevention varies from system to system, apparently a controlled crack-crazing mechanism is favored for many materials.[155,156] This, in turn, arises from a degree of incompatibility between the two polymers, and low-interfacial bonding forces.

More recently, lightly crosslinked, suspension-sized particles have been evaluated. Chemically, networks based on polybutadiene,[157] polystyrene,[157] or poly(vinyl chloride)[158] have been suggested. These particles are swellable in the unsaturated polyester–styrene mix, but are substantially insoluble in it. On curing the polyester, an IPN is formed.

8.11. SUMMARY OF APPLICATIONS

The brief examination of the patent and engineering literature in this chapter has shown a wide range of applicability of the IPN concept. The material covered is representative, rather than complete. To date, there are some 75 patents which explore the application of IPNs and closely related materials. While the Annotated Bibliography gives a more complete listing, it is convenient to summarize some of the more interesting materials (see Table 8.18). While some of the material in Table 8.18 has been referred to elsewhere in this chapter, it is convenient to summarize it in one place.

In general, the use of the IPN concept yields better control over phase domain size and extent of molecular mixing. By various mechanisms, processibility is attained, yet the final products are all crosslinked. Because in many cases the phases interpenetrate to develop dual-phase continuity, a new type of bonding is achieved, to hold the two components in juxtaposition.

REFERENCES

1. L. H. Sperling, K. B. Ferguson, J. A. Manson, E. M. Corwin, and D. L. Siegfried, *Macromolecules* **9**, 743 (1976).
2. E. N. Kresge, in *Polymer Blends*, Vol. 2, D. R. Paul and S. Newman, eds., Academic, New York (1978), Chap. 20.
3. F. G. Hutchinson, Br. Pat. 1,239,701 (1971).
4. F. G. Hutchinson, U.S. Pat. 3,700,752 (1972).
5. F. G. Hutchinson, U.S. Pat. 3,868,431 (1975).
6. F. G. Hutchinson, U.S. Pat. 3,859,381 (1975).
7. F. G. Hutchinson, R. G. C. Henbest, and M. K. Leggett, U.S. Pat. 3,886,229 (1975).
8. F. G. Hutchinson, R. G. C. Henbest, and M. K. Leggett, U.S. Pat. 4,062,826 (1977).

9. G. Allen, M. J. Bowden, D. J. Blundell, F. G. Hutchinson, G. M. Jeffs, and J. Vyvoda, *Polymer* **14**, 597 (1973).

10. G. Allen, M. J. Bowden, D. J. Blundell, G. M. Jeffs, and J. Vyvoda, *Polymer* **14**, 604 (1973).

11. G. Allen, M. J. Bowden, G. Lewis, D. J. Blundell, and G. M. Jeffs, *Polymer* **15**, 13 (1974).

12. G. Allen, M. J. Bowden, G. Lewis, D. J. Blundell, G. M. Jeffs, and J. Vyvoda, *Polymer* **15**, 19 (1974).

13. G. Allen, M. J. Bowden, S. M. Todd, D. J. Blundell, G. M. Jeffs, and W. E. A. Davies, *Polymer* **15**, 28 (1974).

14. D. J. Blundell, G. W. Longman, G. D. Wignall, and M. J. Bowden, *Polymer* **15**, 33 (1974).

15. J. Ferrarini, D. M. Longnecker, N. N. Shah, J. Feltzin, and G. G. Greth, 33rd Annual Conference, RP-C Institute (1978), Sec. 9-D, p. 1.

16. J. Ferrarini, J. J. Magrans, and J. A. Reitz III, 34th Annual Conference, RP-C Institute, SPI (1979), Sec. 2-G, p. 1.

17. J. J. Magrans and J. Ferrarini, 34th Annual Conference, RP-C Institute, SPI (1979), Sect. 2-I, p. 1.

18. J. W. Rees, U.S. Pat. 3,264,272 (1966).

19. J. W. Rees, U.S. Pat. 3,404,134 (1968).

20. E. Kuehn, U.S. Pat. 3,954,714 (1976).

21. N. B. Graham, R. G. C. Henbest, and F. G. Hutchinson, U.S. Pat. 3,860,537 (1975).

22. ICI Americas, Inc., Information Sheets ITP-3, ITP-4, ITP-5, 1–79 250 (1979).

23. K. Kircher and R. Pieper, *Kunststoffe* **68**(3), 141 (1978); *German Plast.* **68**(3), 10 (1978).

24. W. K. Fischer, U.S. Pat. 3,806,558 (1974).

25. Uniroyal, Inc., Br. Pat. 1,380,884 (1975).

26. Uniroyal, Inc. Br. Pat. 1,384,261 (1975).

27. Uniroyal, Inc., Br. Pat. 1,384,262 (1975).

28. W. K. Fischer, U.S. Pat. 3,758,643 (1973).

29. W. K. Fischer, U.S. Pat. 3,862,106 (1975).

30. W. K. Fischer, U.S. Pat. 3,835,201 (1974).

31. L. Mullins, *Rubber Develop.* **31**(4), 92 (1978).

32. *Rubber World* **Feb.**, 49 (1973).

33. G. Holden and R. Milkovich, U.S. Pat. 3,265,765 (1966).

34. R. A. Dickie and S. Newman, U.S. Pat. 3,833,682 (1974).

35. R. A. Dickie and S. Newman, U.S. Pat. 3,833,683 (1974).

36. R. A. Dickie and S. Newman, U.S. Pat. 3,833,883 (1974).

37. D. Klempner, H. L. Frisch, and K. C. Frisch, *J. Polym. Sci. A-2* **8**, 921 (1970).

38. K. C. Frisch, H. L. Frisch, and D. Klempner, Ger. Pat. 2,153,987 (1972).

39. L. N. Johnson, *J. Biomed. Mater. Res. Symp.* **1**, 207 (1971).

40. R. L. Bowen, U.S. Pat. 3,066,112 (1962).

41. C. W. Taylor, U.S. Pat. 3,860,556 (1975).

42. Minnesota Mining and Manufacturing Co., Br. Pat. 1,113,723 (1968).

43. W. Schmidt, R. Purrman, and P. Jochum, U.S. Pat. 3,107,427 (1963).

44. D. E. Waller, U.S. Pat. 3,629,187 (1971).

45. S. Rogers, U.S. Pat. 3,808,170 (1974).

46. D. J. O'Sullivan and T. E. Casey, U.S. Pat. 3,931,678 (1976).

47. C. W. Taylor, U.S. Pat. 3,597,389 (1971).

48. D. G. Stoffey and H. L. Lee, Jr., U.S. Pat. 3,755,420 (1973).

49. D. J. Duff, Br. Pat. 1,450,157 (1976).

50. E. Masuhara, N. Tarumi, N. Nakabayashi, M. Baba, S. Tanaka, E. Mochida, U.S. Pat. 3,829,973 (1974).

51. Kulzer & Co., G.m.b.H., Br. Pat. 1,237,785 (1971).
52. L. D. Schadbolt, Br. Pat. 1,278,413 (1972).
53. Mochida Seiyaku Kabushiki Kaisha, Br. Pat. 1,426,901 (1976).
54. W. B. Eames, S. J. O'Neal, and L. A. Rogers, *J. Am. Dent. Assoc.* **92**, 550 (1976).
55. H. H. Chandler, R. L. Bowen, and G. C. Paffenbarger, *J. Am. Dent. Assoc.* **83**, 344 (1971).
56. A. Acharya and E. H. Greener, *J. Dent. Res.* **51**(5), 1362 (1972).
57. J. W. Harlan and S. H. Feairheller, in *Protein Crosslinking*, Part A, M. Friedman, ed., Plenum, New York (1977).
58. E. H. Harris and S. H. Feairheller, *Polym. Eng. Sci.* **17**, 287 (1977).
59. S. H. Feairheller, A. H. Korn, E. H. Harris, E. M. Filchione, and M. M. Taylor, U.S. Pat. 3,843,320 (1974).
60. H. A. Gruber, E. H. Harris, and S. H. Feairheller, *J. Appl. Polym. Sci.* **21**, 3645 (1977).
61. M. M. Taylor, E. H. Harris, and S. H. Feairheller, *Polym. Prepr.* **19**, 618 (1978).
62. A. H. Korn, M. M. Taylor, and S. H. Feairheller, *J. Am. Leather Chem. Assoc.* **68**, 224 (1973).
63. A. H. Korn, S. H. Feairheller, and E. M. Filachione, *J. Am. Leather Chem. Assoc.* **67**, 111 (1972).
64. E. H. Harris, M. M. Taylor, and S. H. Feairheller, *J. Am. Leather Chem. Assoc.* **69**, 182 (1974).
65. Sun Oil Company, Br. Pat. 952,089 (1964).
66. Esso Research and Engineering Company, Br. Pat. 893,540 (1962).
67. Esso Research and Engineering Company, Br. Pat. 934,640 (1963).
68. H. J. Hagenmeyer, Jr. and M. B. Edwards, U.S. Pat. 3,529,037 (1970).
69. H. J. Hagenmeyer, Jr. and M. B. Edwards, *J. Polym. Sci.* **4C**, 731 (1966).
70. S. Davison and W. P. Gergen, U.S. Pat. 4,041,103 (1977).
71. W. P. Gergen and S. Davison, U.S. Pat. 4,101,605 (1978).
72. F. de Candia, G. Romano, and V. Vittoria, *Rheol. Acta* **16**, 95 (1977).
73. F. de Candia, A. Taglialatela, and V. Vittoria, *J. Appl. Polym. Sci.* **20**, 831 (1976).
74. F. de Candia, A. Taglialatela, and V. Vittoria, *J. Appl. Polym. Sci.* **20**, 1449 (1976).
75. V. F. Shumskii, A. S. Dorozhkii, Yu. S. Lipatov, E. V. Lebedev, and I. P. Getmanchuk, *Kolloidn. Zh.* **38**(5), 949 (1976).
76. T. Nishi, T. K. Kwei, and T. T. Wang, *J. Appl. Phys.* **46**, 4157 (1975).
77. D. L. Siegfried, D. A. Thomas, and L. H. Sperling, accepted, *J. Appl. Polym. Sci.*
78. H. Mizumachi, *J. Adhes.* **2**, 292 (1970); *J. Adhes. Soc. Jpn.* **5**(6) (1969).
79. H. Oberst, L. Bohn, and F. Lindhurdt, *Kunststoffe* **51**, 495 (1961).
80. H. Oberst and A. Schommer, *Kunststoffe* **55**, 634 (1965).
81. H. Warson, *The Application of Synthetic Resin Emulsions*, Enst Benn, London (1972), p. 979.
82. G. S. Pisarenko and A. A. Shemegan, *Probl. Prochn.* **4**, (March 8) (1972).
83. G. L. Ball and I. Salyer, *J. Acoust. Soc. Amer.* **39**, 663 (1966).
84. A. L. Eustice, U.S. Pat. 3,636,158 (1972).
85. H. Oberst, *et al.*, U.S. Pat. 3,547,757; 3,547,758, 3,547,759; 3,547,760; 3,547,755; (1970); 3,533,072; 3,554,885 (1971).
86. R. F. Wollek, U.S. Pat. 3,193,049 (1965).
87. J. H. Botsford, U.S. Pat. 3,534,882 (1970).
88. D. Ross, E. E. Ungar, and E. M. Kerwin, Jr., in *Structural Damping*, J. E. Ruzicka, ed., ASME, New York (1959).
89. R. Plunkett, in *Structural Damping*, J. E. Ruzicka, ed., ASME, New York (1959).

90. E. E. Ungar, in *Noise and Vibration Control*, L. L. Beranek, ed., McGraw-Hill, New York (1971), Chap. 14.
91. T. P. Yin, T. J. Kelly, and J. E. Barry, *J. Eng. Ind. Series B, Trans, ASME*, **89** Part 4, 773 (1967).
92. D. J. Williams, *Polymer Science and Engineering*, Prentice-Hall, Englewood Cliffs, New Jersey (1971).
93. F. Rodriguez, *Principles of Polymer Systems*, McGraw-Hill, New York (1970).
94. J. H. Aklonis, W. J. MacKnight, and M. Shen, *Introduction to Polymer Viscoelasticity*, Wiley-Interscience, New York (1972).
95. G. E. Molau, eds., *Colloidal and Morphological Behavior of Block and Graft Copolymers*, Plenum, New York (1971).
96. N. A. J. Platzer, Chmn., *Multicomponent Polymer Systems*, Advances in Chemistry Series No. 99. American Chemical Society, Washington, D.C. (1971).
97. H. Keskkula, ed., *Polymer Modification of Rubbers and Plastics, J. Appl. Polym. Sci., Appl. Polym. Symp.* No. 7, Interscience, New York (1968).
98. M. Matsuo, *Jpn. Plast.* **2**, (July 6) (1968).
99. S. Manabe, R. Murakami, and M. Takayanagi, *Mem. Fac. Eng. Kyushu Univ.* **28**, 295 (1969).
100. L. H. Sperling, T. W. Chiu, R. G. Gramlich, and D. A. Thomas, *J. Paint Technol.* **46**, 47 (1974).
101. J. A. Grates, D. A. Thomas, E. C. Hickey, and L. H. Sperling, *J. Appl. Polym. Sci.* **19**, 1731 (1975).
102. L. H. Sperling and D. A. Thomas, U.S. Pat. 3,833,404 (1974).
103. I. Skeist, ed., *Handbook of Adhesives*, Reinhold, New York (1962).
104. N. J. L. Megson, *Phenolic Resin Chemistry*, Academic, New York (1958).
105. L. Berend, U.S. Pat 952,724 (1910).
106. L. H. Baekeland, U.S. Pat. 1,111,284 (1914).
107. J. McGavack, U.S. Pat. 1,555,131 (1925).
108. J. McGavack, U.S. Pat. 1,640,363; 1,640,364 (1927).
109. J. W. Aylsworth, U.S. Pat. 1,111,284 (1914).
110. H. Frood, Br. Pat. 176,405 (1920).
111. C. Ellis, *The Chemistry of Synthetic Resins*, Reinhold, New York (1935).
112. F. J. Groten and R. J. Reid, U.S. Pat. 2,581,926 (1952).
113. F. J. Groten and R. J. Reid, U.S. Pat. 2,459,739 (1949).
114. T. Tori, Jpn. Pat. 4,340 (1952); *Chem. Abstr.* **47**, 10902f (1952).
115. P. O. Tawney, J. R. Little, and P. Viohl, *Ind. Eng. Chem.* **51**, 937 (1959).
116. P. O. Tawney and J. R. Little, U.S. Pat. 2,701,895 (1955).
117. Armstrong Cork Co., Br. Pat. 784,565 (1957).
118. J. R. Geigy A.-G., Swiss Pat. 165,713 (1934): *Chem. Abstr.* **28**, 2946 (1934).
119. I. G. Farbenind. A.-G., Br. Pat. 375,320 (1931).
120. B. G. Bufkin and J. R. Grave, *J. Coatings Technol.* **50**(2), 65 (1978).
121. D. H. Solomon, *The Chemistry of Organic Film Formers*, Robert E. Krieger, Huntington, New York (1977), Chap. 4.
122. A. J. Yu and R. E. Gallagher, U.S. Pat. 3,944,631 (1976).
123. C. F. Ryan, U.S. Pat. 3,678,133 (1972).
124. K. Nakatsuka, F. Ide, and R. Handa, U.S. Pat. 3,502,604 (1970).
125. C. F. Ryan and R. J. Crochowski, U.S. Pat. 3,426,101 (1969).
126. A. J. Spilner, U.S. Pat. 3,681,475 (1972).
127. B. Vollmert, U.S. Pat. 3,055,859 (1962).
128. O. B. Johnson and S. S. Labana, U.S. Pat. 3,659,003 (1972).

129. F. Ide, K. Kishida, and A. Hasegawa, Ger. Offen. Pat. 2,619,922 (1976).

130. J. R. Grave and B. G. Bufkin, *J. Coatings Technol.* **51**(2), 34 (1979).

131. Y. Amagi, M. Ohya, Z. Shiiki, and H. Yusa, U.S. Pat. 3,775,514 (1973).

132. General Electric Company, Br. Pat. 1,003,975 (1965).

133. Matsushita Denko Kabushiki Kaisha, Br. Pat. 1,185,665 (1970).

134. Celanese Coatings Company, Br. Pat. 1,205,682 (1970).

135. R. D. Hibelink and G. H. Peters, U.S. Pat. 3,657,379 (1972).

136. W. H. Parriss and R. Orr, Br. Pat. 786,102 (1957).

137. Ciba, Ltd. Br. Pat. 1,136,260 (1968).

138. Standard Oil Co., Br. Pat. 1,100,542 (1968).

139. Westinghouse Electric International Co., Br. Pat. 794,541 (1958).

140. Naamlooze Vennootschap de Bataafsche Petroleum Maatschappij, Br. Pat. 736,457 (1955).

141. Shell Internationale Research Maatschappij N.V., Br. Pat. 1,247,116 (1971).

142. J. M. Hawkins, Br. Pat. 1,197,794 (1970).

143. P. Mendoyanis, U.S. Pat. 3,316,324 (1967).

144. D. C. Prevorsek and S. M. Aharoni, U.S. Pat. 4,055,606 (1977).

145. K. C. Frisch, H. L. Frisch, and D. Klempner, Ger. Offen. Pat. 2,153,987 (1972).

146. W. J. McKillip and C. N. Impola, U.S. Pat. 3,396,210 (1968).

147. J. J. Falcetta, G. D. Friends, and G. C. C. Niu, Ger. Offen. Pat. 2,518,904 (1975).

148. H. A. Clark, U.S. Pat. 3,527,842 (1970).

149. R. A. Jacobson, U.S. Pat. 2,422,550 (1947).

150. Natural Rubber Producers Research Association, Technical Information Sheet No. 9 Revised (1977).

151. E. H. Andrews and D. T. Turner, *J. Appl. Polym. Sci.* **3**, 366 (1960).

152. G. F. Bloomfield and P. McL. Swift, J. Appl. Chem. **5**, 609 (1955).

153. T. D. Pendle, in *Block and Graft Copolymerization*, Vol. 1, R. J. Ceresa, ed., Wiley, New York (1973).

154. Natural Rubber Producers Research Association Technical Information Sheet L14 (1977).

155. K. E. Atkins, in *Polymer Blends*, Vol. 2, Dr. R. Paul and S. Newman, eds., Academic, New York (1978), Chap. 23.

156. V. A. Pattison, R. R. Hindersinn, and W. T. Schwartz, *J. Appl. Polym. Sci.* **19**, 3045 (1975).

157. D. R. Stevenson, U.S. Pat. 4,048,257 (1977).

158. V. F. G. Cooke and D. H. Thorpe, U.S. Pat. 4,125,702 (1978).

ANNOTATED
BIBLIOGRAPHY

These references are restricted to the more important IPN papers and patents, so that not all references cited in the chapter references and Suggested Reading sections are included.

G. Akovali, K. Biliyar, and M. Shen, "Gradient Polymers by Diffusion Polymerization," *J. Appl. Polym. Sci.* **20**, 2419 (1976). Gradient IPNs mechanical behavior.

G. Allen, M. J. Bowden, D. J. Blundell, F. G. Hutchinson, G. M. Jeffs, and J. Vyvoda, "Composites Formed by Interstitial Polymerization of Vinyl Monomers in Polyurethane Elastomers: 1. Preparation and Mechanical Properties of Methyl Methacrylate Based Composites," *Polymer* **14**(12), 597 (1973). Polyurethane/poly(methyl methacrylate) semi–SINs. Interstitial composites. Synthesis and properties. Morphology. A series of six papers by Allen and coworkers. Other papers are in *Polymer* **14**, 605 (1973); **15**, 13, 19, 28, 33 (1974).

Y. Amagi, M. Ohya, Z. Shiiki, and H. Yusa, "PVC Blended with Cross-linked Graft Copolymers of Styrene and Methyl Methacrylate onto a Butadiene Polymer," U.S. Pat. 3,775,514 (1973). MBS resins in latex form. Triple IPN in latex form. Clear, impact–resistant PVC.

Armstrong Cork Company, "Improvements in or Relating to Adhesive Compositions," Br. Pat. 784,565 (1957). NBR-sulfur/phenol-aldehyde-epoxy SINs. High peel strength adhesives.

H. E. Bair, M. Matsuo, W. A. Salmon, and T. K. Kwei, "Radiation-Cross-Linked Poly(vinyl chloride). Phase Studies," *Macromolecules* **5**, 114 (1972). PVC/TEGDM semi-IPNs or ABCs (B crosslinked) grafted, crosslinked PVC morphology of phase separation.

F. P. Baldwin and I. J. Gardner, in "Chemistry and Properties of Crosslinked Polymers," S. S. Labana, ed., Academic, New York (1977). "Graft Curing with Modified Butyl Rubber" and "AB Crosslinked Copolymers of Butyl Rubber with Polystyrene." Crosslinking of the polystyrene. Physical and mechanical Properties. Swelling and extraction.

C. H. Bamford, G. C. Eastmond, and D. Whittle, "Newtork Properties: 2. A Broadline NMR Study of Molecular Motions in Some Multicomponent Crosslinked Polymers," *Polymer* **16**, 377 (1975). Poly(vinyl trichloroacetate)-poly(methyl methacrylate). NMR studies AB-crosslinked polymers.

J. H. Barrett and D. H. Clemens, "Method of Decolorizing Sugar Solutions with Hybrid Ion Exchange Resins," U.S. Pat. 3,966,489 (1976). IPN ion exchange resin. Macroreticular polymer network I. High crosslink levels.

H. A. J. Battaerd, "The Significance of Incompatible Polymer Systems," *J. Polym. Sci. Polym. Symp.* **49**, 149 (1975). Reviews polymer blends and IPNs.

P. Bauer, J. Hennig, and G. Schreyer, "Dynamisch-mechanische Untersuchen an Polymerisatgemischen aus Polyäthylacrylat und Polymethylmethacrylat," *Angew. Makromol. Chem.* **11**, 145 (1970). Blends, grafts, and semi-IPNs of PEA and PMMA. Mechanical behavior.

A. Belkebir-Mrani, J. E. Herz, and P. Rempp, "Unidirected Compression Measurements on Swollen Polystyrene Model Networks," *Makromol. Chem.* **178**(2), 485 (1977). Effect of crosslink functionality in styrene block copolymers and IPN-type network formation.

G. P. Belonovskaya, L. S. Adrianova, Zh. D. Chernova, L. A. Korotneva, and B. A. Dolgoplosk, "New Type of Mutually Penetrating Polymeric Reticular Systems," *Dokl. Akad. Nauk SSSR* **212**(3), 615 (1973). IPN based on isocyanate polymers.

G. P. Belonovskaya, J. D. Chernova, L. A. Korotneva, L. S. Andriarova, L. S. Andriarova, B. A. Dolgoplosk, S. K. Zakharov, Yu. N. Sazanov, K. K. Kalninsh, L. M. Kaljuzhnaya, and M. F. Lebedeva, "Interpenetrating Polymer Networks Based on Diisocyanates and Polar Monomers," *Eur. Polym. J.* **12**(11), 817–823 (1976). Rubber, synthetic poly(propylene sulfide)–poly(tolylene diisocyanate) compatible interpenetrating networks. Polyisocyanate: network, tensile, strength.

G. C. Berry and M. Dror, "Modification of Polyurethanes by Interpenetrating Polymer Network Formation with Hydrogels," *Am. Chem. Soc. Div. Org. Coat. Plast. Chem. Pap.* **38**(1), 465 (1978). Polyether–urethane-urea block copolymers with crosslinked HEMA, NVP, or acrylamide. IPNs and gradient IPNs for biomedical purposes. Strength, water swellability, and good blood compatibility.

W. J. Blank, "Rubbery Polymeric Mixtures Comprising a Maleinized Type Oil," U.S. Pat. 3,719,623 (1973). Semi-IPN or grafted IPN. Triglyceride oil and melamine derivative/styrene-maleic anhydride half ester.

B. A. Bolto, "Process of Making Amphoteric Polymeric Compositions from Snake-Cage Resins," U.S. Pat. 3,875,085 (1975). Anionic/cationic IPNs in semi-I form. Ion exchange resins: weak acid, weak base. Phase separation synthesis. Improvement on Sirotherm process.

M. Braden, "A New Use for Powdered Elastomers," *Plast. Rubber* **1**(6), 241 (1976). IPN linings for dentures. Poly(ethoxyethyl methacrylate)/SBR.

B. G. Bufkin and J. R. Grave, "Survey of the Applications, Properties, and Technology of Crosslinked Emulsions V," *J. Coatings Tech.* **50**(12), 65 (1978). Reviews interstitially crosslinked coatings. Phenol-formaldehyde dispersed in water, etc. Sealants, adhesives, architectural coatings.

A. L. Bull and G. Holden, "The Use of Thermoplastic Rubbers in Blends with Other Plastics," *J. Elastomers Plast.* **9**(7), 281 (1977). Thermoplastic IPNs. Block Copolymers blended with polypropylene. Block copolymers blended with polyethylene.

F. De Candia, G. Romano, and V. Vittoria, "Structure–Property Relationships in Some Composite Systems Visoelasticity," *Rheol. Acta* **16**, 95 (1977). Polybutadiene/methacrylic acid or magnesium methacrylate, simultaneously crosslinked and polymerized. Chemical/Physical IPN. Dynamic mechanical behavior and DSC studies.

F. De Candia, A. Taglialatela, and V. Vittoria, "Structure–Property Relationships in Crosslinked Networks from cis-1, 4-Polybutadiene and Methacrylic Acid. Swelling Behavior," *J. Appl. Polym. Sci.* **20**, 831 (1976). Simultaneous crosslinking of polybutadiene and polymerization of methacrylic acid. Swelling studied via Mooney–Rivlin equation.

F. De Candia, A. Taglialatela and V. Vittoria, "Structure–Property Relationships in Some Composite Systems. Deformation Mechanism," *J. Appl. Polym. Sci.* **20**, 1449 (1976). Simultaneous crosslinking of polybutadiene with polymerization of methacrylic acid or magnesium acrylate. Photoelastic study of hysteresis cycles.

B. E. Causton, "A Study of the Morphology of Denture Base Acrylics Using a Viscoelastic Technique," *J. Dental Res.* **53**(3), 1074 (1974). IPNs as denture-base materials.

Celanese Coatings Company, "Resin Compositions," Br. Pat. 1,205,682 (1970). Unsaturated polyester-styrene/epoxy SIN.

Zh. D. Chernova, K. Kalnins, and G. P. Belonovskaya, "IR–Spectroscopic Study of Some Kinetic Characteristics of the Formation of Interpenetrating Polymer Network Systems," *Vysokomol. Soedin. Ser. B* **19**(1) 61 (1977). IPNs of TDI/polypropylene sulfide.

A. Christou, "Reliability Aspects of Moisture and Ionic Contamination Diffusion Through Hybrid Encapsulants," in *Proceedings of the Technical Program—International Microelectronics Conference* (1978) p. 237, Industrial Scientific Conference Management, Inc., New York. Electric insulators and dielectrics: Silicone/epoxy-polyurethane interpenetrating networks, moisture, and ion diffusion through potting compounds.

Ciba, Ltd. "Curable Filled Resin Compositions," Br. Pat. 1,136,260 (1968). Polyester-styrene/cured aminoplast compositions electrical insulation.

H. A. Clark, "Pressure Sensitive Adhesive Made from Siloxane Resins," U.S. Pat. 3,527,842 (1970). A blend of three PDMS linear resins. Polymer I has $-OH$ groups, polymer II has $-CH=CH_2$, polymer III has $-H$. Si$-H$ crosslinks Si$-CH=CH_2$. Further heating crosslinks polymer I.

V. F. G. Cooke and D. H. Thorpe, "Additive for Low Profile Polymerizable Unsaturated Polyester Molding Composition," U.S. Pat. 4,125,702 (1978). Lightly crosslinked poly(vinyl chloride)-suspension-sized particles dispersed in unsaturated polyester-styrene.

L. S. Corley, P. Kubisa, and O. Vogl, "Haloaldehyde Polymers. V. Polymer Blends Involving Chloral Polymers," *Polym. J.* **9**(1), 47–59 (1977). Blends with polystyrene and poly(meth)acrylates. Crosslinking agents for styrene in presence of polychloral. Crosslinking agents for Me methacrylate in presence of polychloral. IPNs of polychloral/polystyrene and poly(meth)acrylates. Crosslinking agents for polystyrene and poly(Me methacrylate) in blends with polychloral. Interpenetrating network formation.

A. J. Curtius, M. J. Covitch, D. A. Thomas, and L. H. Sperling, "Polybutadiene/Polystyrene Interpenetrating Polymer Networks," *Polym. Eng. Sci.* **12**(2), 101 (1972). Polybutadiene/Polystyrene Network. Interpenetrating polymer network. Impact resistance and glass transition studies.

H. Czarczynska and W. Trochimczuk, "Changes the Structure of Chloromethylated Polyethylene Modified with Styrene and Divinylbenzene Foils due to Friedel–Crafts Catalysts," *J. Polym. Sci.* **47C**, 111 (1974). Semi-II IPNs and IPNs. Chloromethylated structures as points of crosslinking. Effects of $AlCl_3$, $ZnCl_2$, and $SnCl_4$ as catalysts. Optical microscopy of phase separation.

H. Czarczynska and W. Trochimczuk, "Modification of Polyethylene Foil with Methacrylic Acid," Paper P-13 at the Fourth Bratislava IUPAC International Conference on Modified Polymers, their Preparation and Properties, July 1975. Semi-II IPNs using DVB. Selective ion exchange membranes.

A. Damusis and K. C. Frisch, "Flexibilized Polyester Resin Compositions Containing Polyurethanes Modified with Vinyl Monomers and Process Therefore," U.S. Pat. 3,448,172 (1969). Polyester/polyurethane grafted semi-SINs.

S. Davidson and W. P. Gergen, "Blends of Certain Hydrogenated Block Copolymers," U.S. Pat. 4,041,103. Thermoplastic IPNs of the block/crystal type. SEBS/polyamides.

N. Devia-Manjarres, "Synthesis and Characterization of Simultaneous Interpenetrating Networks Based on Castor Oil Elastomers and Polystyrene," *Diss. Abstr. Int. B* **39**(8), 3972 (1979). Castor oil–polyester/PS SIN. Synthesis, morphology, mechanical behavior. Ph.D thesis.

N. Devia-Manjarres, A. Conde, G. Yenwo, J. Pulido, J. A. Manson, and L. H. Sperling, "Castor Oil Based Interpenetrating Polymer Networks. II. Synthesis and Properties of Emulsion Polymerized Products," *Polym. Eng. Sci.* **17**(5), 294 (1977). Sulfur-vulcanized, castor oil/polystyrene latex-type IPN's. Polystyrene/sulfur-vulcanized poly(sodium ricinoleate) blends.

N. Devia-Manjarres, J. A. Manson, L. H. Sperling, and A. Conde, "Simultaneous Interpenetrating Networks Based on Castor Oil Polyesters and Polystyrene," *Polym. Eng. Sci.* **18**(3), 200 (1978). Castor oil–polyester SIN with polystyrene. Polymer morphology of castor oil-polyester/styrene SINs.

N. Devia, J. A. Manson, L. H. Sperling, and A. Conde, "Simultaneous Interpenetrating Networks Based on Castor Oil Elastomers and Polystyrene. V. Behavioral Trends and Analysis," *J. Appl. Polym. Sci.* **24**, 569 (1979). Castor oil–polyester/polystyrene SINs. Morphology and mechanical behavior.

N. Devia-Manjarres, G. Yenwo, J. Pulido, J. A. Manson, A. Conde, and L. H. Sperling, "Castor Oil Based Interpenetrating Polymer Networks: Synthesis and Properties of Emulsion-Polymerized Products," *AIChE Symp. Ser.* **73**(170), 133 (1977). Sulfur-vulcanized/polystyrene IPN. Impact resistance. Sodium ricinoleate as emulsifying agent for polymerization of polystyrene.

N. Devia, J. A. Manson, and L. H. Sperling, "Simultaneous Interpenetrating Networks Based on Castor Oil Elastomers and Polystyrene. III. Morphology and Glass Transition Behavior," *Polym. Eng. Sci.* **19**(12), 869 (1979). Castor oil–polyester/styrene SINs. Electron microscopy and T_g. Studies in phase domain formation.

N. Devia, J. A. Manson, and L. H. Sperling, "Simultaneous Interpenetrating Networks Based on Castor Oil Elastomers and Polystyrene. IV. Stress–Strain and Impact Loading Behavior," *Polym. Eng. Sci.* **19**(12), 878 (1979). Castor oil-polyester/PS SINs. Mechanical behavior. Impact resistant plastics.

N. Devia, J. A. Manson, L. H. Sperling, and A. Conde, "Simultaneous Interpenetrating Networks Based on Castor Oil Elastomers and Polystyrene. 2. Synthesis and Systems Characteristics," *Macromolecules* **12**(3), 360 (1979). Synthesis and Processing of castor oil–polyester/PS SINs. Rubber-toughened plastics and reinforced elastomers.

R. A. Dickie, M. F. Cheung, and S. Newman, "Heterogeneous Polymer–Polymer Composites. II. Preparation and Properties of Model Systems," *J. Appl. Polym. Sci.* **17**, 65 (1973). Latex Semi-IPNs. Poly(methyl methacrylate)/poly(*n*-butyl acrylate). Synthesis, morphology, mechanical behavior.

R. A. Dickie and S. Newman, "Rubber-Modified Thermosets and Processes," U.S. Pat. 3,833,682 (1974). Semi-I and IPN latexes with reactive shells. Graded composition latexes containing rubber cores. Thermoset epoxies, etc. containing reactive latexes. One of three closely related patents. See U.S. Pat. 3,833, 683 (1974) and 3,856,883 (1974).

B. N. Dinzburg, V. P. Popova, V. G. Dynunina, and A. E. Chalykh, "Effect of the Structure of Styrene–Butadiene Copolymers on the Properties of Resins in Admixture with Rubbers," *Kolloidn. Zh.* **38**(2), 338 (1976). SBR(High S)/SBS block copolymer (80% S)/PS, sulfur vulcanized. Improved tearing resistance.

A. A. Donatelli, "Morphology and Mechanical Behavior of Poly(butadiene-costyrene)/Polystyrene Interpenetrating Polymer Networks," *Diss. Abstr. Int. B.* **36**(5), 2369 (1975). SBR/PS IPNs. Ph.D. thesis.

A. A. Donatelli, D. A. Thomas, and L. H. Sperling, "Poly(butadiene-styrene)/Polystyrene IPN's, Semi-IPN's and Graft Copolymers, Staining Behavior and Morphology," *Polym. Sci. Technol.* **4**, 375 (1974); reprinted in *Recent Advances in Polymer Blends, Grafts, and Blocks*, L. H. Sperling, ed., Plenum, New York (1974). SBR/PS IPNs and semi-IPNs. Morphology and staining behavior.

A. A. Donatelli, L. H. Sperling, and D. A. Thomas, "Interpenetrating Polymer Networks Based on SBR/PS. 1. Control of Morphology by Level of Crosslinking," *Macromolecules* **9**(4), 671 (1976). SBR/Polystyrene IPN. Morphology and crosslinking level.

A. A. Donatelli, L. H. Sperling, and D. A. Thomas, "Interpenetrating Polymer Networks Based on SBR/PS. 2. Influence of Synthetic Detail and Morphology on Mechanical Behavior," *Macromolecules* **9**(4), 676 (1976). SBR/Polystyrene IPN. Mechanical Properties.

A. A. Donatelli, L. H. Sperling, and D. A. Thomas, "A Semiempirical Derivation of Phase Domain Size in Interpenetrating Polymer Networks," *J. Appl. Polym. Sci.* **21**(5), 1189 (1977). Equations for phase domain size in IPNs and semi-I IPNs. Effect of crosslink density, composition, interfacial tension.

G. C. Eastmond and E. G. Smith, "Some Morphological Aspects of AB Crosslinked Polymers," *Polymer* **17**, 367 (1976). Transmission electron microscopy. Synthesis of AB-crosslinked polymers.

G. C. Eastmond and D. G. Phillips, "Macroscopic Phase Separation in Multicomponent Polymer Homopolymer Blends: General Considerations Based on Studies of AB-Crosslinked Polymers," *Polymer* **20**, 1501 (1979). Morphology, especially onion rings.

Esso Research and Engineering Company, "Polyethylene Diluted Polypropylene," Br. Pat. 893,540 (1962). Physical or thermoplastic crystalline/crystalline IPNs. Polyethylene/polypropylene blends. Improved low-temperature properties.

Esso Research and Engineering Company, "Improved Polypropylene–Polyethylene Blends," Br. Pat. 934,640 (1963). Physical or thermoplastic crystalline/crystalline IPNs. Polyethylene/polypropylene blends. Improved low-temperature properties.

J. J. Falcetta, G. D. Friends, G. C. C. Niu, "Solid Article Formed from a Polymer Network with Simultaneous Interpenetration," Fr. Demande 2365606 (1978). Moldings, containing interpenetrating networks of crosslinked hydrophobic siloxane chains, as contact lenses. SINs of hydroxy ethyl methacrylate/PDMS. Soft contact lenses.

J. J. Falcetta, G. D. Friends, and G. C. C. Niu, "Molding From an Interpenetrating Network Polymer," Ger. Offen. 2,518,904 (1975). Soft contact lenses of HEMA/PDMS SINs.

J. A. Faucher and M. R. Rosen, "Shaped Article for Conditioning Hair, a Blend of Water-Soluble and Water-Insoluble Polymers with Interpenetrating Networks," U.S. Pat. 4,018,729 (1977). Polymer blend, hair conditioning combs of (polycaprolactone blend, hair conditioning combs. Hair preparations, conditioners; water insoluble/water soluble polymer blends and IPN-related materials. Combs and shaped articles.

S. H. Feairheller, A. H. Korn, E. H. Harris, Jr., E. M. Filachione, and M. M. Taylor, "Graft Polymerization of Vinyl Monomers onto Chrome-Tanned Hides and Skins," U.S. Pat. 3,843,320 (1974). Semi-I IPNs of vinyl monomers grafted onto leather. Emulsion process for transporting the monomer into the leather. Mechanical properties.

R. P. Fellmann, W. H. Stass, and D. R. Kory, "Impact-Resistant Polymer Mass," Ger. Offen. 2,748,751 (1978). Interpenetrating networks with poly(Me methacrylate). Co-continuous interpenetrating networks of chains. Ethylene vinyl acetate copolymer/poly(Me methacrylate) blends.

J. Ferrarini, D. M. Longnecker, N. N. Shah, J. Feltzin, and G. G. Greth, "Development of a Unique Method for the Preparation of High Quality SMC," 33rd Annual Technical Conference, R-P/C Institute, SPI, P. 1, Section 9-D (1978). ITP^{TM}, interpenetrating thickening process. Sheet molding compounds based on SINs. Polyurethane/styrene-polyester.

J. Ferrarini, J. J. Magrans, and J. A. Retiz III, "New Resins for High Strength SMC," 34th Annual Technical Conference, R-P/C Institute, SPI, P. 1, Section 2-G (1979). Sheet molding compounds, ITP^{TM} processed SMC. Polyurethane/styrene-polyester.

W. K. Fischer (see Uniroyal, Inc.), "Dynamically Partially Cured Thermoplastic Blend of Monoolefin Copolymer Rubber and Polyolefin Plastic," U.S. Pat. 3,806,558 (1974). EPM or EPDM partly cured, blended with polypropylene or polyethylene. Thermoplastic IPNs.

H. L. Frisch, R. Foreman, R. Schwartz, H. Yoon, D. Klempner, and K. C. Frisch, "Barrier and Surface Properties of Polyurethane–Epoxy Interpenetrating Polymer Networks. II," *Polym. Eng. Sci.* **19**(4), 294 (1979). Polyurethane/epoxy SINs. Contact angles of drops of methanol mixtures on polyurethane/epoxy interpenetrating network films. Transmission of vapors in polyurethane/epoxy SINs.

H. L. Frisch and K. C. Frisch, "Polyurethane–Epoxy Interpenetrating Polymer Networks— Barrier and Surface Properties," *Prog. Org. Coat.* **7**, 107 (1979). Epoxy/Polyurethane SINs. Lap-shear, critical surface tension, and permeability studies.

H. L. Frisch, K. C. Frisch, and D. Klempner, "Glass Transition of Topologically Interpenetrating Polymer Networks," *Polym. Sci.* **14**(9), 646 (1974). Topological IPN. Glass Transitions. Polyurethane-based SIN.

H. L. Frisch, K. C. Frisch, and D. Klempner, "Tangled Polymers," *Chem. Tech.* (Amer. Chem. Soc.), **7**(3), 188 (1977). Review of SINs, IPNs, and IENs.

H. L. Frisch, K. C. Frisch, and D. Klempner, "Interpenetrating Polymer Networks," *Mod. Plast.* **54**(4), 76 (1977). Review of IPN literature.

H. L. Frisch, K. C. Frisch, and D. Klempner, "Examining the Properties of Interpenetrating Polymer Networks," *Mod. Plast.* **54**(5), 84 (1977). Polymer morphology of SINs. Review.

H. L. Frisch, K. C. Frisch, and D. Klempner, "Interpenetrating Polymer Networks, Chemical Properties of Crosslinked Polymers," in *Chemistry and Properties of Crosslinked Polymers*, S. S. Labana, ed., Academic, New York (1977). Review of Polyurethane SINs.

H. L. Frisch and D. Klempner, "Topological Isomerism and Macromolecules," *Adv. Macromol. Chem.* **2**, 149 (1970). Macromolecular catenanes. Review of IPN structures IENs.

H. L. Frisch, D. Klempner, and K. C. Frisch, "Topologically Interpenetrating Elastomeric Networks," *J. Polym. Sci. Part B* **7**(11), 775, (1969). IENs based on polyurethanes.

K. C. Frisch, "Topologically Interpenetrating Polymer Networks," *Pure Appl. Chem.* **43**, 229 (1975). Topological Interpenetration. Stress/strain behavior, tensile strength, T_g. Polyurethane SINs.

K. C. Frisch, H. L. Frisch, and D. Klempner, "Kunstoff und Verfahrung zu seiner Herstellung," Ger. Offen. 2,153,987 (1972). IEN and SIN polyurethane compositions.

K. C. Frisch, D. Klempner, T. Antczak, and H. L. Frisch, "Stress–Strain Properties of Polyurethane–Polyacrylate Interpenetrating Polymer Networks," **18**(3), 683 (1974). Polyacrylate/Polyurethane SINs.

K. C. Frisch, D. Klempner, H. L. Frisch, and H. Ghiradella, "Topologically Interpenetrating Polymer Networks," *Polym. Sci. Technol.* **4**, 395 (1974); reprinted in *Recent Advances in Polymer Blends, Blocks, and Grafts*, L. H. Sperling, ed., Plenum, New York (1974). PU/Polyacrylate SINs.

K. C. Frisch, D. Klempner, S. Migdal, and H. L. Frisch, "Polyurethane–Polyacrylate Interpenetrating Polymer Networks. I," *J. Polym. Sci. Polym. Chem. Ed.* **12**(4), 885 (1974). Polyurethane/Polyacrylate SINs. Impact strength. Heat resistance. Tensile strength.

K. C. Frisch, D. Klempner, S. Migdal, H. L. Frisch, and H. Ghiradella, "Morphology of a Polyurethane–Polyacrylate Interpenetrating Polymer Network," *Polym. Eng. Sci.* **14**(1), 76 (1974). Polyacrylate/polyurethane SINs. Morphology and glass transitions.

K. C. Frisch, D. Klempner, S. K. Mukherjee, and H. L. Frisch, "Stress–Strain Properties and Thermal Resistance of Polyurethane–Polyepoxide, Interpenetrating Polymer Networks," *J. Appl. Polym. Sci.* **18**(3), 689 (1974). Polyurethane/Epoxy SIN. Tensile strength. Heat resistance.

General Electric Company, "Resinous Insulating Composition," Br. Pat. 1,003,975 (1965). Polyester-styrene/epoxy-chlorinated anhydride SIN. Solid dielectric and protective coating for electrical apparatus.

W. P. Gergen and S. Davison, "Thermoplastic polyester/Block Copolymer Blend," U.S. Pat. 4,101,605 (1978). Blends of ABA block copolymers and crystalling polyesters. Physical IPNs with continuous interlocking networks. Selectively hydrogenated block copolymers. Thermoplastic IPNs of the block/crystal type.

N. B. Graham, R. G. C. Henbest, and F. G. Hutchinson, "Process for Preparing a Foamed Shaped Article," U.S. Pat. 3,860,537 (1975). Polyurethane/Polyester IPN-related SIN synthesis. Foaming process. Unsaturated polyesters and styrene.

J. A. Grates, D. A. Thomas, E. C. Hickey, and L. H. Sperling, "Noise and Vibration Damping with Latex Interpenetrating Polymer Networks," *J. Appl. Polym. Sci.* **19**(6), 1731 (1975). Latex IPNs for sound damping. Methacrylic/acrylic compositions.

J. R. Grave and B. G. Bufkin, "Survey of the Applications, Properties, and Technology of Crosslinked Emulsions. VI," *J. Coatings Tech.* **51**(2), 34 (1979). Review of latex IPNs and related materials.

R. T. Greer, B. H. Vale, and R. L. Knoll, "Hydrogel coatings and Impregnations in Silastic, Dacron and Polyethylene," *Scanning Electron Microsc.* **1**, 633, (1978). Coating process of poly(hydroxyethyl methacrylate) hydrogels of polyester fibers by gamma irradiation. Surface morphology of coated fibers.

C. C. Gryte and H. P. Gregor, "Interpolymer Ultrafiltration Membranes Prepared from Poly(vinylidene Fluoride) and Poly(1-vinyl-3-methylimidazolium Iodide)," *J. Macromol. Sci.-Phys.* **B15**(2), 183 (1978). Poly(vinyl imidazole)/poly(vinylidene fluoride)/epoxy semi-IPN. Ion exchange membrane.

S. C. Hargest, J. A. Manson, and L. H. Sperling, "Morphological Features of Anionic/Cationic Interpenetrating Polymer Networks," AIChE Preprints, 87th National Meeting, Boston, Massachusetts, August 1979, paper No. 14e. PS/P(4-VP) anionic/cationic IPNs. IPNs before and after ionic introduction. T_g, morphology, swelling behavior.

E. H. Harris and S. H. Feairheller, "Crosslinking Systems in the Graft Polymerization of Chromium-Tanned Collagen," *Polym. Eng. Sci.* **17**, 287 (1977). IPN formation with tanned leather as network I. Crosslinked acrylics, etc. serve as network II. Mechanical and physical behavior. Swelling behavior.

E. H. Harris, M. M. Taylor, and S. H. Feairheller, "Graft Polymerization III, Some Properties of the Leather Obtained from the Graft Polymerization of Vinyl Monomers onto Chrome-Tanned Nigerian Hairsheep," *J. Am. Leather Chem. Assoc.* **69**, 182 (1974). IPNs based on leather and crosslinked acrylics. Physical and mechanical behavior. Swelling and extraction studies.

M. J. Hatch, "Composite Ion Exchange Resin Bodies," U.S. Pat. 3,041,292 (1962). Sulfonated polystyrene/poly(acrylic acid) semi-I compositions.

M. J. Hatch, "Method of Making Composite Ion Exchange Resin Bodies," U.S. Pat. 3,205,184 (1965). Semi-I IPN compositions. "Snake-cage" resins. Ion exchange materials.

M. J. Hatch, "Process of Making Composite Ion Exchange Resin Bodies," U.S. Pat. 3,332,890 (1967). Semi-IPN compositions. Ion exchange resins.

M. J. Hatch, "Thermally Reversible, Amphoteric Ion Exchange Resins Consisting of Cross-linked Microbeads Embedded in Crosslinked Matrix of Opposite Exchange Group Type," U.S. Pat. 3,957,698 (1976). IPN ion exchange resin. Weak acid and weak base.

J. M. Hawkins, "Epoxy Resin Adhesive Compositions," Br. Pat. 1,197,794 (1970). Epoxy/polyurethane semi-IPNs.

R. D. Hibelink and G. H. Peters, "Intercrossing Resin/Curing Agent Adhesive Systems," U.S. Pat. 3,657,739 (1972). Epoxy/Polyester IPNs.

F. G. Hutchinson, "Polymeric Shaped Articles," Br. Pat. 1,239,701 (1971). Polyurethane/PMMA SINs and semi-SINs.

F. G. Hutchinson, "Gel Polymerized Polyurethane Precursors and Vinyl Monomers," U.S. Pat. 3,700,752 (1972). Semi–SINs and SINs. Crosslinked polyurethane and vinyl monomers. Gel-polymerization process.

F. G. Hutchinson, "Crosslinked Polyurethanes for Polyurethane Precursers and Vinyl Monomers," U.S. Pat. 3,859,381 (1975). SIN's. Elastomeric polyurethanes and vinyl polymers.

F. G. Hutchinson, "Elastomer Production," U.S. Pat. 3,868,431 (1975). SINs based on crosslinked elastomeric polyurethane and ethylenically unsaturated monomers, i.e., MMA + DEGDM. Reinforced elastomers.

F. G. Hutchinson, R. G. C. Henbest, and M. K. Leggett, "Shaped Polymeric Articles," U.S. Pat. 3,886,229 (1975). Semi-SINs. Linear polyurethanes plus polyester–styrene resin.

F. G. Hutchinson, R. G. C. Henbest, and M. K. Leggett, "Polymeric Shaped Articles," U.S. Pat. 4,062,826 (1977). SIN's, grafted. Plastic polyurethanes based on 4:4'-diphenyl-methane diisocyanate. Styrene-unsaturated polyester. Gel polymerization. Shaping of partly gel-polymerized articles.

V. Huelck, D. A. Thomas, and L. H. Sperling, "Interpenetrating Polymer Networks of Poly(ethyl acrylate) and Poly(styrene-*co*-methyl methacrylate). I. Morphology via Electron Microscopy," *Macromolecules* **5**(4), 340 (1972). Polyacrylate/Polystyrene IPNs. Polymethacrylate/Polyacrylate IPNs. Morphology via electron microscopy.

V. Huelck, D. A. Thomas and L. H. Sperling, "Interpenetrating Polymer Networks of Poly(ethyl acrylate) and Poly(styrene-*co*-methyl methacrylate). II. Physical and Mechanical Behavior," *Macromolecules* **5**(4), 348 (1972). PEA/PS, PEA/PMMA sequential IPNs. Mechanical behavior, T_g.

F. Ide, K. Kishida, and A. Hasegawa, "Multilayered Polymer Materials," Ger. Offen. 2,619,922 (1976). Multilayered latex IPNs. Overcoating five times. MMA and BA in various proportions, allyl methacrylate as crosslinker.

L. N. Johnson, "Composite Resins as a Dental Restorative Material," *J. Biomed. Mater. Res. Symp.* **1**, 207 (1971). History of use of polymers in tooth fillings. Description of composite materials for tooth fillings. Description of linear PMMA/crosslinked PMMA semi-II IPNs.

O. B. Johnson and S. S. Labana, "Thermoset Molding Powders from Hydroxy-Functional Graded Elastomer Particles and Monoblocked Diisocyanate and Molded Article," U.S. Pat. 3,659,003 (1972). Acrylic/methacrylic IPNs. Latex-based, rubber-toughened plastics.

L. M. Kalyuzhnaya, L. S. Andrianova, Zh.D. Chernova, Glp. Belonovskaya, and S. Ya. Frenkel, "Sorption of Vapors of a Polar Monomer and Solvents by a Polyisocyanate Network," *Vysokomol. Soedin. Ser. B.* **19**(2), 143 (1977). Sorption of benzene by interpenetrating networks with poly(tolylene diisocyanate). Poly(tolylene diisocyanate)/polypropylene sulfide IPNs.

D. S. Kaplan, "Structure–Property Relationships in Copolymers to Composites: Molecular Interpretation of the Glass Transition Phenomenon," *J. Appl. Polym. Sci.* **20**, 2615 (1976). Reviews morpholgy of IPNs.

D. Kaplan and N. W. Tschoegl, "Mechanical and Optical Properties of Polyurethane–Polystyrene Two-Phase Polymers," *Polym. Eng. Sci.* **15**, 343 (1975). Polyurethane/polystyrene SINs, Mechanical and optical behavior.

M. Kapuscinski and H. P. Schreiber, "Effect of Dispersion on Flow and Mechanical Properties of Polymer Blends," *Polym. Eng. Sci.* **19**, 900 (1979). Thermoplastic IPNs of polyethylene and ethylene–propylene–diene elastomer.

S. C. Kim, D. Klempner, K. C. Frisch, H. L. Frisch, and H. Ghiradella, "Polyurethane–poly(methyl methacrylate) interpenetrating polymer networks," *Am. Chem. Soc. Div. Org. Coat. Plast. Chem. Pap.* **35**(2), 31 (1975). PU/PMMA SINs.

S. C. Kim, D. Klempner, K. C. Frisch, and H. L. Frisch, "Polyurethane Interpenetrating Polymer Networks II. Density and Glass Transition Behavior of Polyurethane–Poly(methyl methacrylate) and Polyurethane–Polystyrene IPNs," Macromolecules 9(2), 263 (1976). Polyurethane/Polymethacrylate SIN Polystyrene/Polyurethane SIN. Glass transition and density studies.

S. C. Kim, D. Klempner, K. C. Frisch, and H. L. Frisch, "Polyurethane Interpenetrating Polymer Networks. 3. Visoelastic Properties of Polyurethane-poly(methyl methacrylate): Interpenetrating Polymer Networks," Macromolecules 10(6), 1187 (1977). Mechanical loss. Viscoelasticity of poly(methyl methacrylate)/polyurethane SINs.

S. C. Kim, D. Klempner, K. C. Frisch, and H. L. Frisch, "Polyurethane Interpenetrating Polymer Networks. 4. Volume Resistivity Behavior of Polyurethane–poly(methyl methacrylate) Interpenetrating Polymer Networks," Macromolecules 10(6), 1191 (1977). Volume resistivity behavior of SINs. PU/PMMA SINs.

S. C. Kim, D. Klempner, K. C. Frisch, and H. L. Frisch, "Polyurethane Interpenetrating Networks. V. Engineering Properties of Polyurethane–poly(methyl methacrylate) IPNs," J. Appl. Polym. Sci. 21(5), 1289 (1977). PU/PMMA SINs. Tensile strength.

S. C. Kim, D. Klempner, K. C. Frisch, H. L. Frisch, and H. Ghiradella, "Polyurethane–Polystyrene Interpenetrating Polymer Networks," Polym. Eng. Sci. 15(5), 339 (1975). Polystyrene/polyurethane SINs. Phase Separation. T_g and mechanical behavior.

S. C. Kim, D. Klempner, K. C. Frisch, W. Radigan, and H. L. Frisch, "Polyurethane Interpenetrating Polymer Networks. I. Synthesis and Morphology of Polyurethane–Poly(methyl methacrylate) Interpenetrating Polymer Networks," Macromolecules 9(2), (1976). PU/PMMA SINs. Morphology and compatibility studies.

S. C. Kim, "Interpenetrating Polymer Networks," Hwahak Konghak 14(1), 17 (1976). Review of IPNs.

K. Kircher, "Kombinationswerkstoffe aus Polyurethan und Vinyl-polymeren," Angew. Makromol. Chem. 76/77, 241 (1979). Polymer blends and IPNs based on polyurethanes and poly(methyl methacrylate). SINs. Electron microscopy and mechanical behavior. Impact-resistant plastics.

K. Kircher, and G. Menges, "Polystyrene–Polyurethane Interpenetrating Networks—New Substance for Liquid Injection Molding," Plast. Eng. 32(10), 37 (1976). Polystyrene/Polyurethane SINs.

K. Kircher and R. Pieper, "Polyurethane–Polymethyl methacrylate Copolymers," Kunststoffe 68(3), 141 (1978). SINs of PU and PMMA are described. Moldings were prepared. Swellability flexural strength, impact strength, and surface hardness were determined.

D. Klempner, "Topologically Interpenetrating Elastomeric Networks," Diss. Abstr. Int. B 31(6), 3321 (1970); Univ. Microfilms, Ann Arbor, Mich., Order No. 70–25, 444. Study of IENs based on PU. Ph.D. thesis.

D. Klempner, "Polymer Networks with Mutual Penetration," University of Detroit, Polymer Institute, Detroit, Michigan; Angew. Chem. 90(2), 104, (1978). Review of SINs and IPNs.

D. Klempner and H. L. Frisch, "Thermal Analysis of Polyacrylate Poly(urethane urea) Interpenetrating Polymer Networks," J. Polym. Sci. Part B 8(7), 525 (1970). PU/polyacrylate IENs.

D. Klempner and K. C. Frisch, "Polyurethane Interpenetrating Networks," Adv. Urethane Sci. Technol. 3, 14 (1974). Review of polyurethane SINs.

D. Klempner, H. L. Frisch, and K. C. Frisch, "Topologically Interpenetrating Elastomeric Networks," J. Polym. Sci. Part A-2 8, 921 (1970). IENs based on polyurethane.

D. Klempner, H. L. Frisch, and K. C. Frisch, "Topologically Interpenetrating Polymeric Networks," J. Elastoplast. 3 (January), 2 (1971). PU/Polyacrylate and other IENs.

D. Klempner, K. C. Frisch, and H. L. Frisch, "Nomenclature of Interpenetrating Polymer Networks," *J. Elastoplast.* **5** (October), 196 (1973). Nomenclature of IPNs and related materials.

D. Klempner, H. K. Yoon, K. C. Frisch, and H. L. Frisch, "Polyurethane–Polyacrylate Pseudo-Interpenetrating Networks Chemical Properties of Crosslinked Polymers," in *Chemistry and Properties of Crosslinked Polymers*, S. S. Labana, ed., Academic, New York (1977). SINs of urethane rubbers. Morphological and mechanical properties. Glass temperature, transitions and polymer morphology of urethane rubber/acrylic copolymer SINs.

V. I. Klenin, M. Yu. Prozorova, L. S. Adrianova, Yu. V. Brestkin, G. P. Belonovskaya, and S. Ya. Frenkel, "Study of the Structure of Mutually Penetrating Polymeric Systems Using Turbidity Spectrum and X-Ray Diffraction Methods," *Vysokomol. Soedin. Ser. A.* **19**(5), 1138 (1977). IPNs of poly(tolylene diisocyanate. Poly(alkylene sulfide/polydiisocyanate IPNs.

B. Kolarz, "Porous Carboxylic Cation Exchangers," *Polymer* **2** (1975). Cation Exchanger Styrene Divinylbenzene. Macroporous Carboxy Cation Exchanger. IPNs based on PS.

B. N. Kolarz, "Ion Exchangers XIX. Some Properties of the Carboxylic Cation Exchangers Obtained by Intermesh Polymerization of Methacrylic Acid into Styrene and Divinylbenzene Porous Copolymers," *J. Polym. Sci.* **47C**, 197 (1974). PS/PMA IPNs. Both polymers crosslinked with DVB. Effect of porosity of network I. Effect of crosslink density of network I. Ion exchange properties.

B. N. Kolarz, "Some Properties of Carboxylic Cation Exchangers Produced by the Intermeshing Polymerization of Methacrylic Acid in Porous Styrene–Divinylbenzene Copolymers," *Akad. Nauk BSSR Inst. Obshch. Neorg. Khim.*, Minsk, USSR 184 (1975). Poly(methacrylic acid)/PS IPN-type cation exchangers.

B. N. Kolarz, "Interpenetrating Polymer Networks Part II. Poly(methacrylic acid *co*-divinyl benzene)–Poly(styrene-*co*-divinylbenzene)," Report No. 8, "Instytut Technologii Organicznej I Tworzyw Sztucznych" (1979). IPNs of Polystyrene/Poly(methacrylic acid). Electron microscopy showing inhomogeneities within each network and between networks. Theory of interpenetration vs. void filling. Ion exchange properties.

J. V. Koleske, C. J. Whitworth, Jr. and R. D. Lundberg, "Nylon Polymers Blended with Cyclic Ester Polymers," U.S. Pat. 3,781,381 (1973) Thermoplastic IPNs of polyesters and nylons.

A. H. Korn, S. H. Feairheller, and E. M. Filachione, "Graft Polymerization. I. Preliminary Results with Acrylate Esters," *J. Am. Leather Chem. Assoc.* **67**, 111 (1972). Semi-I IPNs based on leather and acrylics. Method of synthesis.

A. H. Korn, M. M. Taylor, and S. H. Feairheller, "Graft Polymerization. II. Factors Affecting the Graft Polymerization of Vinyl Monomers onto Chrome-Tanned Hide Substance," *J. Am. Leather Chemists Assoc.* **68**, 224 (1973). Semi-I IPNs based on leather and acrylics. Discussion of synthesis procedures.

Yu. S. Lipatov, "Role of Interfacial Phenomena in the Formation of Micro and Macro-Heterogeneities in Multicomponent Polymer Systems," *Pure Appl. Chem.* **43**, 273 (1975). Review of blends and filled IPNs. Physical properties, electron microscopy.

Yu. S. Lipatov, V. F. Babich, L. V. Karabanova, N. I. Korzhukh, and L. M. Sergeeva, "Study of Viscoelastic Properties of Mutually Penetrating Polymer Networks," *Dopov. Akad. Nauk Ukr. Rsr. Ser. B* **1**, 39 (1976). PU/PS IPNs. Viscoelastic behavior.

Yu. S. Lipatov, L. V. Karabanova, T. S. Khramova, and L. M. Sergeeva, "Study of Physicochemical Properties of Interpenetrating Polyurethane and Poly(urethane acrylate) Polymer Networks," *Vysokomol. Soedin. Ser. A* **20**(1), 46 (1978). Thermodynamics and Swelling of IPNs and PU-based IPNs.

Yu. S. Lipatov, L. V. Karabonava, and L. M. Sergeeva, "Thermodynamic Study of the Swelling of Filled Mutually Penetrating Polymeric Networks Based on Polyurethane and a Styrene–Divinylbenzene Copolymer," *Vysokomol. Soedin. Ser. A.* **19**(5), 1073 (1977). Swelling and Sorption of filled PU/PS IPNs.

Yu. S. Lipatov, L. V. Karabanova, L. M. Sergeeva, and A. E. Fainerman, "Diffusion of Benzene Vapors into Interpenetrating Polymer Lattices," *Sint. Fiz.-Khim. Polim.* **18**, 63 (1976). Swelling of PS-based IPN system.

Yu. S. Lipatov, L. V. Karabanova, T. S. Khramova, and L. M. Sergeeva, "Study of Physicochemical Properties of Interpenetrating Polyurethane and Poly(urethane acrylate) Polymer Networks," *Vysokomol. Soedin. Ser. A.* **20**(1), 46 (1978). Thermodynamics of sorption. Swelling behavior of PU-based IPNs.

Yu. S. Lipatov, T. S. Khramova, L. M. Sergeeva, and L. V. Karabanova, "Some Properties of Intermediate Regions in Interpenetrating Polymeric Networks," *J. Polym. Sci. Polym. Chem. Ed.* **15**(2), 427 (1977). PU/PS IPNs. Study of interfacial properties.

Yu. S. Lipatov, A. E. Nesterov, L. M. Sergeeva, L. V. Karabanova, and T. D. Ignatova, "Thermodynamics of Mutually Penetrating Polymeric Networks," *Dokl. Akad. Nauk. SSSR* **220**(3), 637 (1975). Polyurethane interpenetrating network thermodynamics. PU/PS IPNs.

Yu. S. Lipatov and L. M. Sergeeva, "Synthesis and Properties of Interpenetrating Networks," *Usp. Khim.* **45**(1), 138 (1976). Review of IPNs.

Yu. S. Lipatov and L. M. Sergeeva, *Interpenetrating Polymeric Networks*, Naukova Dumka, Kiev (1979). Review of IPNs. Interfacial, mechanical, glass transition temperature properties emphasized. Book on IPNs.

Yu. S. Lipatov and L. M. Sergeeva, *Interpenetrating Polymeric Networks*, Naukova Dumka, Kiev (1979). Review of IPNs. Interfacial, mechanical, glass transition temperature properties emphasized. Book on IPNs.

Yu. S. Lipatov, L. M. Sergeeva, L. V. Karabanova, A. E. Nesterov, and T. D. Ignatova, "Thermodynamic and Sorption Properties of Interpenetrating Polymer Networks Based on Polyurethane and a Styrene–Divinylbenzene Copolymer," *Vysokomol. Soedin. Ser. A.* **18**(5), 1025 (1976). Transition layer studies of PU/PS IPNs. IPN heat of mixing.

Yu. S. Lipatov, L. M. Sergeeva, L. V. Mozzhukhina, and N. P. Apukhtina, "Physicochemical Properties of Filled Interpenetrating Networks," *Vysokomol. Soedin. Ser. A* **16**(10), 2290 (1974). PU/PS IPNs. Dielectric loss, sorption isotherms. Effect of filler.

Yu. S. Lipatov, R. A. Veselovskii, and Yu. K. Znachkov, "Characteristics of the Properties of Adhesives Based on Self-Penetrating Polymer Networks," *Dokl. Akad. Nauk SSSR* **238**(1), 174 (1978). PU/Polyester IPN adhesives. Bonding strength.

T. E. Lipatova, V. V. Shilov, N. P. Bazilevskaya, and Yu. S. Lipatov, "The Formation of Interpenetrating Polymeric Networks on the Basis of Oligoester Acrylate, Styrene, and Divinylbenzene by an Anionic Mechanisms," *Br. Polym. J.* **9**(2), 159 (1977). Oligoester acrylate/PS anionically prepared IPNs.

J. D. Lipko, H. F. George, D. A. Thomas, S. C. Hargest, and L. H. Sperling, "Transparent Fluorocarbon-Based Semi-II IPN Elastomers by High-Energy Radiation Methods," *J. Appl. Polym. Sci.* **23**, 2739 (1979). Crosslinking of fluorocarbon elastomers. Synthesis of transparent elastomers. Characterization via modulus and swelling. Semi-II IPNs with grafting.

D. F. Lohr, Jr. and W. Kang, "Blend of High-Vinyl Polybutadiene and Hydroformylated High-Vinyl Polybutadiene," U.S. Pat. 3,928,282 (1975). Blended elastomers simultaneously vulcanized and grafted.

F. Lohse, "Chemical Structure Principles of Thermosetting Plastics," *Prog. Colloid Polym. Sci.* **64**, 1 (1978). (Formerly *Fortschr. der. Kolloide Polym.*) Study of thermosetting plastics, IPNs, and ionomers.

J. E. Lorenz, D. A. Thomas, and L. H. Sperling, "Viscoelastic Properties of Acrylic Latex Interpenetrating Polymer Networks as Broad Temperature Span Vibration Damping Materials," ACS Symp. Ser. 24, *Emulsion Polymerization*, I. Piirma and J. L. Gardon, eds., American Chemical Society, Washington, D.C. (1976). Methacrylic/acrylic IPNs. Controlled temperature range damping compositions. Latex IPNs. Constrained layer damping.

J. J. Magrans and J. Ferrarini, "Unique Electrical and Mechanical Properties of ITP™ SMC," 24th Annual Technical Conference, R-P/C Institute, SPI, P.1, Section 2-E (1979). Sheet molding compounds. ITP™ process. Electrical and Mechanical Properties.

G. C. Martin, E. Enssani, and M. Shen, "The Mechanical Behavior of Gradient Polymers," AIChE Preprints, 87th National Meeting, Boston, Massachusetts, August 1979, paper No. 14d. Gradient IPNs and sequential IPNs. PMMA/poly(2-chloroethyl acrylate). Glass transition and mechanical studies.

E. Masuhara, N. Tarumi, N. Nakabayashi, M. Baba, S. Tanaka, and E. Mochida, "Dental and Surgical Bonding-Filling Material," U.S. Pat. 3,829,973 (1974). Linear PMMA/Cross-linked PMMA semi-II IPNs. Trialkylboron initiators for acrylic systems. Tooth filling applications.

M. Matsuo, T. K. Kwei, D. Klempner, and H. L. Frisch, "Structure Property Relations in Polyacrylate–Poly(urethane-urea) Interpenetrating Polymer Networks," *Polym. Eng. Sci.* **10**(6), 327 (1970). Polyacrylate/polyurethane urea IENs. Polyurethane/urea polyacrylate IENs. Morphology via electron microscopy.

Matsushita Denko Kabushika Kaisha, "Ballast Composition for Electrical Devices," Br. Pat. 1,185,665 (1970). Polyester-styrene/castor oil-TDI SINs.

W. J. McKillip and C. N. Impola, "Compositions made from (A) Isocyanate-Terminated Prepolymers; and (B) Polyesters Prepared from Polyols and α, β, Ethylenically Unsaturated Monocarboxylic Acids," U.S. Pat. 3,396,210 (1968). IPN-type compositions.

P. Mendoyanis, "Thermosetting Compositions Containing a Liquid Rubber Selected from Polysulfide, Polymercaptan, and Chlorinated Polyethylene. Together with an Epoxide and Curing Agent," U.S. Pat. 3,316,324 (1967). SIN compositions of epoxy and polysulfide rubber.

Mochida Seigaku Kabushike Kaisha (Company), "Dental and Surgical Bonding-Filling Materials," Br. Pat. 1,426,901 (1976). PMMA (linear)/PMMA (crosslinked) semi-II IPNs. Tooth filling applications. Trialkylboron initiators for acrylic systems. (See Masuhara *et al.*)

G. Meyer, "Interpenetrating Polymer Networks," *Rev. Gen. Caoutch. Plast.* **54**(573), 99 (1977). Polyurethane-based SINs.

G. C. Meyer and P. Y. Mehrenberger, "Polyester–Polyurethane Interpenetrating Networks," *Eur. Polym. J.* **13**(5), 383 (1977). Urethane-based SINs.

J. R. Millar, "Interpenetrating Polymer Networks—Styrene–Divinylbenzene Copolymers with Two and Three Interpenetrating Networks, and Their Sulphonates," *J. Chem. Soc.*, 1311 (1960). Synthesis of IPNs. Swelling Behavior of IPNs. vs. crosslink density.

J. R. Millar, "Some Aspects of Ion Exchanger Structure and Synthesis," in *Kunstharz Ionenaustauscher* Akademie Berlin (1970). Sulfonated copolymer with two interpenetrating polymer networks. Ion exchange properties. Swelling properties.

J. R. Millar, D. G. Smith, and W. E. Marr, "Interpenetrating Polymer Networks. Part II. Kinetics and Equilibria in a Sulphonated Secondary Intermeshed Copolymer," *J. Chem. Soc.*, 1789 (1962). IPNs as ion exchange resins. Kinetics of ion exchange. Swelling behavior in water.

Y. Minoura, "Entanglement in crosslinking Polymers: Interpenetrating Polymer Networks," *Kobunshi* **27**(3), 189 (1978). Review of IPNs.

H. Mizumachi and Y. Ogata, "One-Step Synthesis of Viscoelastic Properties of Interpenetrating Polymer Networks Comprised of Polyurethane and Poly(methylmethacrylate)," *Nippon Setchaku Kyokai Shi* **12**(1) (1976). PU/PMMA SINs.

L. Mullins, "Advances in Thermoplastic Natural Rubber," Rubber Dev. **31**(4), 92 (1978). Review of thermoplastic IPNs.

Naamlooze Vennootschap de Bataafsche Petroleum Maatschappij, "Mixtures of Glycidyl Polyethers and Butadiene Copolymers," Br. Pat. 736,457 (1955). Semi-SINs and SINs based on epoxy-amine/NBR. High strength adhesion to metal.

T. Nakamura, "Preparation Techniques for Interpenetrating Polymer Networks of Polyurethanes," *Purasuchikkusu* **27**(7), 50 (1976). Review of polyurethane-based IPNs.

K. Nakatsuka, F. Ide, and R. Handa, "Impact Resistant Resin Compositions and Method of Production Thereof," U.S. Pat. 3,502,604 (1970). Latex semi-Is and latex blends. Impact-resistant plastics. Alkyl acrylates form seed latex. Styrene, methyl methacrylate, etc. from shell.

E. A. Neubauer, D. A. Thomas, and L. H. Sperling, "Effect of Decrosslinking and Annealing on Interpenetrating Polymer Networks Prepared from Poly(ethyl acrylate)/polystyrene Combinations," *Polymer* **19**(2), 188 (1978). Decrosslinking of PEP/PS IPNs. Hydrolysis of Acrylic acid anhydride. Morphology and physical properties.

T. Nishi, T. K. Kwei, and T. T. Wang, "Physical Properties of Poly (vinyl chloride)–Copolyester Thermoplastic Elastomer Mixtures," *J. Appl. Phys.* **46**, 4157 (1975). PVC/copolyester thermoplastic elastomer thermoplastic IPN. Modified Hytrel materials.

B. V. Ozerkovskii, Yu. B. Kalmykov, U. G. Gafurov, and V. P. Roshchupkin, "Thermodynamic and Kinetic Factors in Formation of the Morphology of Polymeric Composites," *Vysokomol. Soedin* **A19**(7), 1437 (1977). Solution blends, semi-SINs and SINs prepared from butyl rubber and acrylics. Optical microscopy, dielectric loss factor, and interfacial tension were studied.

W. H. Parriss and R. Orr, "Improvements in Casting Resins," Br. Pat. 786,102 (1957). Styrene–polyester AB-crosslinked copolymer as network I. Epoxy Resin as network II. SIN compositions.

V. A. Patterson, R. R. Hindersinn, and W. T. Schwartz, "Mechanism of Low-Profile Behavior in Single-Phase Unsaturated Polyester Systems," *J. Appl. Polym. Sci.* **19**, 3045 (1975). Semi-IPNs of unsaturated polyester–styrene and poly(vinyl acetate). Low profile behavior. Craze-cracking behavior to control shrinkage.

N. A. Peppas, C. T. Reinhart, G. S. Sekhon, and R. Sorenson, "Diffusion of Macromolecules Through Model Hydrophilic Networks," AIChE Preprints, 87th National Meeting, Boston, Massachusetts, August 1979, Paper No. 14f. Semi-IPNs. Diffusion of a linear polymer through a crosslinked one.

P. Predecki, "A Method for Hydron Impregnation of Silicone Rubber," *J. Biomed. Mater. Res.* **8**, 487 (1974). IPNs of silicone rubber and poly(hydroxyethyl methacrylate). Implant materials, particularly arteriovenous shunts.

D. C. Prevorsek and S. M. Aharoni, "Novel Copolyester–Polyepoxide Compositions," U.S. Pat. 4,055,606 (1977). Copolyester/epoxy semi-II IPNs. Epoxy composed of high, intermediate, and low MW. High green strength adhesive.

D. C. Prevorsek and G. E. R. Lamb, "Blends of Polyamids, Polyesters, and Polyolefins Containing Minor Amounts of Elastomers," U.S. Pat. 3,546,319 (1970). Thermoplastic IPNs of the crystal/crystal and crystal/glass type.

J. E. Pulido-Florez, G. M. Yenwo, L. H. Sperling, and J. A. Manson, "Mechanical Properties of New Polymers Made from Castor Oil and Styrene," *Rev. Univ. Ind. Santander Invest.* **7**(7), 35 (1977). Castor oil-urethane/PS IPNs. Mechanical behavior.

A. Rembaum and C. J. Wallace, "Membrane Consisting of Polyquaternary Amine Ion Exchange Polymer Network Interpenetrating the Chains of Thermoplastic Matrix Polymer," U.S. Pat. 4,119,581 (1978). Anionic ion exchange membranes. Semi-IPNs.

Rohm and Haas Co. "Schlagfeste Polymermasse," Ger. Offen. 2,748,751 (1976). PMMA/Poly(ethylene-*co*-vinylacetate) blends exhibiting co-continuous phases.

V. F. Rosovizky, M. Ilavsky, J. Hrouz, K. Dusek, and Yu. S. Lipatov, "Viscoelastic Behavior of Interpenetrating Networks of Polyurethane and Polyurethane Acrylate," *J. Appl. Polym. Sci.* **24**, 1007 (1979). SINs of polyurethane diacrylate and polyurethane. Swelling, mechanical, and optical properties. Interfacial layer interpretation.

V. P. Roshchupkin, I. S. Kochneva, and B. V. Ozerkovsky, "A Structural Kinetic Method of the Formation of Crosslinked Polymeric Compositions," *Vysokomol. Soedin.* **A20**(10), 2252 (1978). Kinetics of chemical and microsyneresis processes in nonylacrylate/butyl rubber undergoing simultaneous vulcanization.

C. F. Ryan and R. J. Crochowski, "Acrylic Modifiers which Impart Impact Resistance and Transparency to Vinyl Chloride Polymers," U.S. Pat. 3,426,101 (1969). Three-layered latex IPN: Poly(butyl acrylate), network I. Polystyrene, network II. Poly(methyl methacrylate), linear polymer III. Latex dispersed in poly(vinyl chloride) or copolymers.

I. Sakurada, "Interpenetrating Polymer Networks of Two Different Polymers," *Kobunshi Kako* **23**(9), 406 (1974). Review of Polyurethane IPNs.

P. R. Scarito and L. H. Sperling, "Effect of Grafting on Phase Volume Fraction, Composition, and Mechanical Behavior, Epoxy/Poly(*n*-butyl acrylate) Simultaneous Interpenetrating Networks," *Polym. Eng. Sci.* **19**, 297 (1979). Epoxy/poly(*n*-butyl acrylate) SINs. Synthesis, morphology, mechanical behavior. Molecular mixing, effect on T_g in each phase.

J. M. Scohy, G. E. Cremeans, and M. Luttinger, "Copolymer Binder Resins for Traffic Paints," U.S. Pat. 3,928,266 (1975). Linear polystyrenes dissolved in alkyd paint formulations. Semi-II IPNs.

Shell Internationale Research Maatschappij N.V., "Process for Producing High-Impact Thermosetting Compositions," Br. Pat. 1,247,116 (1971). Elastomer-epoxy IPNs. Elastomer latexes, crosslinked.

M. Shen and H. Kawai, "Properties and Structure of Polymeric Alloys," *AIChE J.* **24**(1), 1 (1978). Review of block copolymers and IPNs. Gradient IPNs. Thermodynamics of mixing.

K. Shibayama and Y. Suzuki, "Viscoelastic Properties of Multiple Network Polymers. IV. Copolymers of Styrene and Divinylbenzene," *Rubber Chem. Tech.* **40**, 476 (1967). Homo-IPNs of polystyrene and polystyrene. Swelling and mechanical behavior.

V. V. Shilov and T. E. Lipatov, "Structure of the Interpenetrating Networks Obtained by Anionic Polymerization," *Vysokomol. Soedin. Ser. A.* **20**(1), 62 (1978). Anionic IPNs based on PS and oligoester acrylates.

V. V. Shilov, L. V. Karabanova, Yu. S. Lipatov, and L. M. Sergeeva, "Structure of Interpenetrating Polymeric Networks Based on Polyurethane and Polyurethane Acrylate," *Vysokomol. Soedin. Ser A* **20**(3), 643 (1978). Acrylate-terminated polyurethane-based IPNs.

V. V. Shilov, Yu. S. Lipatov, L. V. Karabanova, and L. M. Sergeeva, "Phase Separation in the Interpenetrating Polymeric Networks on the Basis of Polyurethane and Polyurethane Acrylates," *J. Polym. Sci. Polym. Chem. Ed.* **17**, 3083 (1979). Small-angle X-ray study. Thickness of transition layer, diffuseness of phase boundary, degree of miscibility determined.

V. F. Shumskii, A. S. Dorozhkii, Yu. S. Lipatov, E. V. Lebedev, and I. P. Getmanchuk, "Rheological Properties of Mixtures of Crystallizable Polymers," *Kolloidn. Zh.* **38**(5), 949 (1976). PE/POM crystalline/crystalline thermoplastic IPN. Rheological behavior.

D. L. Siegfried, J. A. Manson, and L. H. Sperling, "Viscoelastic Behavior and Phase Domain Formation in Millar Interpenetrating Polymer Networks of Polystrene," *J. Polym. Sci. Polym. Phys. Ed.* **16**(40), 583 (1978). PS/PS homo-IPNs. Visoelastic and morphological behavior.

K. J. Smith, Jr. and R. J. Gaylord, "Non-Gaussian Elasticity of Composite and Interpenetrating Networks," *J. Polym. Sci. Part A-2* **10**(2), 283 (1972). Equation for the modulus of SIN-type IPNs. Theory of rubber elasticity for composite systems.

G. S. Solt, "Improvements Relating to the Production of Ion-Exchange Resins," Br. Pat. 728,508 (1955). Anionic/cationic IPNs. Ion-exchange resins. Suspension polymerization.

L. H. Sperling, "Generation of Novel Polymer Blend, Graft and IPN Structures Through the Application of Group Theory Concepts," *Polym. Sci. Technol.* **4**, *Recent Advances in Polymer Blends, Grafts, and Blocks*, L. H. Sperling, ed., Plenum, New York (1974). Group theory concepts applied to polymer blends, blocks, grafts, and IPNs.

L. H. Sperling, "Application of Group Theory Concepts to Polymer Blends, Grafts, and IPN's," *Am. Chem. Soc. Div. Org. Coat. Plast. Chem. Pap.* **34**(2), 282 (1974). Group theory concepts applied to polymer blends, grafts, blocks, and IPNs.

L. H. Sperling, "Application of Group Theory Concepts to Polymer Blends, Grafts, and IPN's in *Toughness and Brittleness of Plastics*, Advances in Chemistry Series No. 154, R. D. Deanin and A. M. Crugnola, eds. American Chemical Society, Washington, D.C. (1976). Review of IPNs and the application of group theory concepts for nomenclatures.

L. H. Sperling, "Interpenetrating Polymer Networks," *Encycl. Polym. Sci. Technol. Suppl.* **1**, 288 (1976), Wiley, New York, H. F. Mark and N. M. Bikales, eds. Review of IPNs, IENs, and SINs.

L. H. Sperling, "Mechanical Behavior of Polymer Blends, Blocks, Grafts, and Interpenetrating Networks," *J. Polym. Sci. Polym. Symp.* **60**, 175 (1977). Review of polymer blends, grafts, blocks, and IPNs. Mechanical behavior and morphology.

L. H. Sperling, "Isomeric Graft Copolymers and Interpenetrating Polymer Networks. Current Status of Nomenclature Schemes," in *Chemistry and Properties of Crosslinked Polymers*, S. S. Labana, ed., Academic, New York (1977). Group theory concepts applied to polymer blends, grafts, blocks, and IPNs. Nomenclature scheme.

L. H. Sperling, "Interpenetrating Polymer Networks and Related Materials," *J. Polym. Sci. Macromol Rev.* **12**, 141 (1977). Review of IPNs in relation to polymer blends and grafts.

L. H. Sperling, "Interpenetrating Polymer Networks and Related Materials," *Polym. News* **4**(5), 206 (1978). Review of IPNs.

L. H. Sperling, T.-W. Chiu, C. P. Hartman, and D. A. Thomas, "Latex Interpenetrating Polymer Networks," *Polym. Prepr. Am. Chem. Soc. Div. Polym. Chem.* **13**(2), 705 (1972). Damping properties of latex IPNs based on methacrylic/acrylic compositions.

L. H. Sperling, T.-W. Chiu, R. G. Gramlich, and D. A. Thomas, "Latex IPN Coating System for Damping Noise and Vibrations over a Broad Temperature Range," U.S. Nat. Tech. Inform. Serv., Ad Rep. No. 765511/1. Latex IPNs as damping materials.

L. H. Sperling, T.-W. Chiu, R. G. Gramlich, and D. A. Thomas, "Synthesis and Behavior of Prototype Silent Paint," *J. Paint Technol.* **46**, 47 (1974). Constrained layer damping. Latex IPNs. Methacrylic/acrylic compositions.

L. H. Sperling, T. W. Chiu, and D. A. Thomas, "Glass-Transition Behavior of Latex-Interpenetrating Polymer Networks Based on Methacrylic–Acrylic Pairs," *J. Appl. Polym. Sci.* **17**(8), 2443 (1973). Polyacrylate/polymethacrylate latex IPNs. Glass transition and damping behavior.

L. H. Sperling and E. M. Corwin, "A Proposed Generalized Nomenclature Scheme for Multipolymer and Multimonomer Systems," in *Multiphase Polymers*, S. L. Cooper and G. M. Estes, eds., Advances in Chemistry Series No. 176, American Chemical Society,

Washington, D.C. (1979). Nomenclature for polymer blends, blocks, grafts, and IPNs. Ring theory application.

L. H. Sperling and K. B. Ferguson, "Isomeric Graft Copolymers and Interpenetrating Polymer Networks. Possible Arrangements and Nomenclature," *Macromolecules* **8**(6), 69 (1975). Graft copolymer and IPN nomenclature scheme. Application of group theory concepts.

L. H. Sperling, K. B. Ferguson, J. A. Manson, E. M. Corwin, and D. L. Siegfried, "Isomeric Graft Copolymers and Interpenetrating Polymer Networks. Theory and Experiment," *Macromolecules* **9**(5), 743 (1970). Application of ring theory concepts to IPN nomenclature. Decrosslinking of PS as example.

L. H. Sperling, V. A. Forlenza, and J. A. Manson, "Interpenetrating Polymer Networks as Piezodialysis Membranes," *J. Polym. Sci. Polym. Lett. Ed.* **13**(12), 713 (1975). Piezodialysis membrane. Cationic/anionic IPNs. PS, sulfonated/poly(vinyl pyridine), quaternized sequential IPNs.

L. H. Sperling and D. W. Friedman, "Synthesis and Mechanical Behavior of Interpenetrating Polymer Networks: Poly(ethyl acrylate) and Polystyrene," *J. Polym. Sci. A-2* **7**, 425 (1969). Synthesis of sequential IPNs. Modulus-temperature behavior. Modulus-composition behavior.

L. H. Sperling, H. F. George, V. Huelck, and D. A. Thomas, "Viscoelastic Behavior of Interpenetrating Polymer Networks: Poly(ethyl acrylate)–Poly(methyl methacrylate)," *J. Appl. Polym. Sci.* **14**, 2815 (1970). Creep behavior of sequential IPNs. Stress relaxation. Master curves.

L. H. Sperling, J. A. Grates, J. E. Lorenz, and D. A. Thomas, "Noise damping with methacrylate/acrylate latex interpenetrating polymer networks," *Polym. Prepr. Am. Chem. Soc. Div. Polym. Chem.* **16**(1), 274 (1975). Methacrylic/acrylic latex IPNs. Broad-temperature noise-damping characteristics.

L. H. Sperling, V. Huelck, and D. A. Thomas, "Morphology and Mechanical Behavior of Interpenetrating Polymer Networks," in *Polymer Networks: Structure Mechanics and Properties*, A. J. Chompff and S. Newman, eds., Plenum, New York (1971). PEA/PS and PEA/PMMA sequential IPNs. Glass transitions.

L. H. Sperling, J. A. Manson, G. M. Yenwo, A. Conde, and N. Devia, "Castor Oil Based Interpenetrating Polymer Networks," *Polym. Prepr. Am. Chem. Soc. Div. Polym. Chem.* **16**(2), 604 (1975). Castor oil–urethane/PS IPNs. Glass transitions and morphology.

L. H. Sperling, J. A. Manson, G. M. Yenwo, N. Devia-Manjarres, J. Pulido, and A. Conde, "Novel Plastics and Elastomers from Castor Oil Based IPN's: A Review of an International Program," in *Polymer Alloys*, D. Klempner and K. C. Frisch, eds., Plenum, New York (1977). Castor oil–urethane/polystyrene sequential IPNs. Synthesis, morphology, and mechanical behavior. Fatigue behavior.

L. H. Sperling and E. N. Mihalakis, "Swelling Behavior of Interpenetrating Polymer Networks Containing Ionic Groups," *J. Appl. Polym. Sci.* **17**(12), 3813 (1973). Ammonium acrylate/PS IPNs. Swelling behavior.

L. H. Sperling and H. D. Sarge, III, "Joined and Sequential Interpenetrating Polymer Networks Based on Poly(dimethylsiloxane)," *J. Appl. Polym. Sci.* **16**(11), 3041 (1972). PDMS/PS and PDMS/PMMA sequential IPNs and ABCPs.

L. H. Sperling and D. A. Thomas, "Vibration or Sound Damping Coating for Vibratory Structures," U.S. Pat. 3,833,404 (1974). Latex IPNs. Constrained layer damping.

L. H. Sperling, D. A. Thomas, J. E. Lorenz, and E. J. Nagel, "Synthesis and Behavior of Poly(vinyl chloride)-Based Latex Interpenetrating Polymer Networks," *J. Appl. Polym. Sci.* **19**(8), 2225 (1975). PVC/Nitrile rubber latex IPNs. Glass temperatures.

A. L. Spliner, "Thermoplastic Molding Compositions," U.S. Pat. 3,681,475 (1972). Acrylic latexes, some IPNs. Rubber-toughened PMMA compositions.

Standard Oil Co. "Reinforced Synthetic Plastic Composition," Br. Pat. 1,100,542 (1968). PVC/polyester-styrene plus crosslinker semi-II compositions. Reinforced plastics.

J. J. P. Staudinger and H. M. Hutchinson, "Process for the Production of Strain-Free Masses from Crosslinked Styrene-Type Polymers," U.S. Pat. 2,539, 377 (1951). Homo-IPNs and semi-I IPNs based on polystyrene.

H. L. Stephens, R. W. Roberts, T. F. Reed, and R. J. Murphy, "Starch-Elastomer Masterbaches," *Ind. Eng. Prod. Res. Develop.* **10**(1), 84 (1971). Semi-IPNs prepared from rubber latex and aqueous polymer solution. Zinc starch xanthate or starch xanthide solutions. SBR or NBR latexes. Mixing of starch and latex, and coagulation. Physical and mechanical properties, oil resistance.

D. R. Stevenson, "Pigmentable Low Shrink Thermosetting Polyesters," U.S. Pat. 4,048,257 (1977). Lightly crosslinked suspension-sized particles dispersed in unsaturated polester–styrene mix to prevent shrinkage.

Sun Oil Company, "Improvements in or Relating to Polypropylene," Br. Pat. 952,089 (1964). Physical or thermoplastic IPN, crystalline/crystalline. Polyethylene/polypropylene blends. Very-high-molecular-weight polyethylene. High-impact strength.

M. M. Taylor, E. H. Harris, and S. H. Feairheller, "Effect of Chain Transfer Agents on the Viscosity Molecular Weight of Methyl Methacrylate Grafted onto Chromium-III Crosslinked Collogen," *Polym. Prepr.* **19**, 618 (1978). Semi-I IPNs based on leather and acrylics. Molecular weight control of graft.

J. L. Thiele and R. E. Cohen, "Synthesis and Characterization of Single-Phase Interpenetrating Polymer Networks," *Polym. Prepr.* **19**(1), 137 (1978). Swelling and modulus behavior Polystyrene/polystyrene IPNs. Swelling equation.

J. L. Thiel and R. E. Cohen, "Synthesis, Characterization, and Viscoelastic Behavior of Single-Phase Interpenetrating Styrene Networks," *Polym. Eng. Sci.* **19**, 284 (1979). Polystyrene/polystyrene homo-IPNs. Swelling equation for single-phase IPNs. Equilibrium swelling studies as a function of crosslink level.

D. A. Thomas, "Morphology Characterization of Multiphase Polymers by Electron Microscopy," *J. Polym. Sci. Polym. Symp.* **60**, 189 (1977). Morphology of polymer blends and IPNs. Electron microscopy techniques.

D. A. Thomas and L. H. Sperling, in *Polymer Blends*, Vol. 2, D. R. Paul and S. Newman, eds., Academic, New York (1978), Chapter II. Review of IPN literature through 1976.

R. E. Touhsaent, "Simultaneous Interpenetrating Polymer Networks of Epoxy and Poly(n-butyl acrylate), Morphology and Mechanical Behavior," *Diss. Abstr. Int. B*, 3864 (1975). **35**(8). Epoxy/P(n-BA) SINs. Simultaneous gelation. Mechanical behavior. Ph.D. thesis.

R. E. Touhsaent, D. A. Thomas, and L. H. Sperling, "Epoxy/acrylic Simultaneous Interpenetrating Networks," *J. Polym. Sci. Polym. Symp.* **46**, 175 (1974). Epoxy/P(n-BA) SINs. Morphology and mechanical behavior.

R. E. Touhsaent, D. A. Thomas, and L. H. Sperling, "Simultaneous Interpenetrating Networks Based on Epoxy/Acrylic Materials," in *Toughness and Brittleness of Plastics*, R. D. Deanin and A. M. Crugnola, Advances in Chemistry Series No. 154, American Chemical Society, Washington, D.C. (1976), Epoxy/P (n-BA) SINs.

W. Trochimczuk, "Changes in Structure of Polyethylene/Styrene divinyl benzene System," presented at the "Structure and Properties of Polymer Networks," Jablonna, Poland, April 1979. Polyethylene/poly(styrene-co-divinyl benzene) semi-II IPNs. Scanning electron microscopy of etched samples. Morphology goes from PS discontinuous to PS continuous as DVB level is increased passed 2%. Polyethylene crystal size is decreased with increasing DVB content.

W. Trochimczuk, "Polyethylene-Poly(styrene-co-divinyl benzene) System. II. Swelling and Specific Volume Measurements," Presented at the IUPAC Sponsored Conference,

"Modification of Polymers," Bratislava, Czechoslovakia, July 1979. Polyethylene/poly(styrene-*co*-divinylbenzene) semi-II IPNs. Swelling regain studies. Specific volume studies.

E. B. Trostyanskaya, A. S. Tevlina, and I. P. Losev, "Polymerization of Monomers in Swelling Copolymers," *Vysokomol. Soedin.* **2**(9), 1413 (1960). Semi-I IPNs. Ion exchange materials. Methacrylic or vinylsulfonic acids in sulfonated crosslinked polystyrene. Methylvinylpyridine in crosslinked vinylbenzyltrimethyl ammonium base.

Uniroyal, Inc. (see W. K. Fischer), "Dynamically Partially Cured Thermoplastic Blend of Monoolefin Copolymer Rubber and Polyolefin Plastic," Br. Pat. 1,380,884 (1975). Dynamic partial cure of EPDM or EPM. Thermoplastic elastomers. Blend with polyethylene or polypropylene. Simultaneous blending and semicuring.

Uniroyal, Inc. (see W. K. Fischer), "Thermoplastic Blend of Partially Cured Monoolefin Copolymer Rubber and Polyolefin Plastic," Br. Pat. 1,384,261 (1975). Partially cured EPDM or EPM blended with polyethylene or polypropylene. Thermoplastic elastomers. Sequential partial cure and blending.

B. Vollmert. "Impact-Resistant Plastic Compositions Comprising Styrene Polymer and a Cross-Linked Acrylic Acid Ester Polymer, and Process for Preparing Same," U.S. Pat. 3,055,859 (1962). Numerous multipolymer grafts and IPNs. Impact-resistant plastics.

W. D. Waters, "Modified Vinyl Ester Resin and Pipe Made Therefrom," U.S. Pat. 3,928,491 (1975). CTBN + epoxy resin + methacrylic acid gives polymer I. Styrene as polymer II. An ABCP composition of I and II.

Westinghouse Electric International Co., "Improvements in or Relating to Resinous Compositions," Br. Pat. 794,541 (1958). Unsaturated polyester-styrene/epoxy SINs. Low-shrinkage resins.

J. M. Widmaier, "Anionic Crosslinking of Polystyrene in the Presence of Free Polymeric Chains," *Makromol. Chem.* **179** (7), 1743 (1978). PS-based semi–IPNs.

E. J. Willinger, "Plastic Laminate Structure Having Elastomer Particles in Alternate Layers," U.S. Pat. 3,620,900 (1971). Polyester–styrene resins, ABCPs, containing rubber in alternate layers. Rubber is linear or crosslinkable.

W. D. Willis, "Covulcanizing Process," U.S. Pat. 3,351,517 (1967). Covulcanizing epihalohydrin polymers with sulfur-curable rubbers.

G. M. Yenwo, "Synthesis, characterization, and Behavior of Interpenetrating Polymer Networks and Solution Graft Copolymers Based on Castor Oil and Polystyrene," *Diss. Abstr. Int. B* **37**(11), 5788, (1977). Castor oil–urethane/PS sequential IPNs. Synthesis, morphology, glass transitions, mechanical properties. Ph.D. thesis.

G. M. Yenwo, J. A. Manson, J. Pulido, L. H. Sperling, A. Conde, and N. Devia, "Castor Oil Based Interpenetrating Polymer Networks, Synthesis and Characterization," *J. Appl. Polym. Sci.* **21**(6), 1531 (1977). Castor oil–urethane/PS sequential IPNs. Physical properties.

G. M. Yenwo, L. H. Sperling, J. A. Manson, and A. Conde, "Castor Oil Based Interpenetrating Polymer Networks. III. Characterization and Morphology," in *Chemistry and Properties of Crosslinked Polymers*, S. S. Labana, ed., Academic, New York (1977). Castor oil–urethane/PS IPNs. Morphology via electron microscopy.

G. M. Yenwo, L. H. Sperling, J. Pulido, J. A. Manson, and A. Conde, "Castor oil Based Interpenetrating Polymer Networks IV. Mechanical Behavior," *Polym. Eng. Sci.* **17**(4), 251 (1977). Castor oil–urethane/PS IPNs. Glass transition and mechanical behavior.

J. K. Yeo, "Controlled Variation of Poly(*n*-Butyl Acrylate)/Polystyrene IPN Morphology and Behavior," Ph.D. thesis, Lehigh University, in preparation.

H. K. Yoon, "Pseudo-Interpenetrating Polymer Networks," Ph.D. thesis, University of Detroit, in preparation.

A. J. Yu and R. E. Gallagher, "Acrylate–Styrene–Acrylonitrile Composition and Method of Making the Same," U.S. Pat. 3,944,631 (1976). Latex IPN of acrylate elastomer/styrene-acrylonitrile. Latex IPN embedded in linear styrene-acrylonitrile. Rubber-toughened, impact-resistant plastics.

INDEX